# ORGANIZATION AND PERFORMANCE OF COTTON SECTORS IN AFRICA

# ORGANIZATION AND PERFORMANCE OF COTTON SECTORS IN AFRICA:
## LEARNING FROM REFORM EXPERIENCE

### CONFERENCE EDITION

**Editors**
   David Tschirley
   Colin Poulton
   Patrick Labaste

**THE WORLD BANK**
**Washington, DC**

# Contents

**TABLES**

**FIGURES**

## BOXES

# Foreword

Cotton is a major source of foreign exchange earnings in more than 15 countries across all regions of Sub-Saharan African (SSA) and a crucial source of cash income for millions of rural people in these countries. The crop is therefore critical in the fight against rural poverty. The World Bank and other development institutions have been and are currently assisting many cotton-exporting countries of SSA to improve their cotton sector performance through projects supporting investment as well as through policy and institutional reform.

Many SSA countries have been implementing or are considering implementing reforms of their cotton industries. The ultimate objective of the reform programs is to strengthen the competitiveness of cotton production, processing, and exports in an increasingly demanding world market and to ensure long-term, sustainable, and equitable growth for these major sectors of many African economies. The reform programs generally entail redefining the role of the state; facilitating greater involvement of the private sector and farmer organizations; ensuring greater competition in input and output markets; improving productivity through research and development, extension, and technology dissemination; and seeking value addition through market development and processing of cotton lint and by-products.

This study was undertaken by the Environmentally and Socially Sustainable Development Department of the Africa Region of the World Bank to fill a perceived gap in knowledge on the lessons to be drawn from nearly two decades of cotton sector reforms in SSA. Recent experience in policy dialogue, particularly with West African countries, shows that very often the analytical points of reference are limited to neighboring countries. At a time when the design of cotton sector reform programs has become extremely complex and potentially risky, stronger and broader analysis, drawing on a broader array of empirical evidence, and reflecting strategically on potential options, would be very useful for policy makers. The lack of such analysis, especially of the reform options available and of their possible implications, partly explains the reluctance of many governments to engage in ambitious restructuring of their cotton sectors. Therefore, the main objective of this study is to provide an in-depth and comparative analysis of the reforms that have been implemented by SSA cotton sectors since 1990, and from there, to establish links between reforms and observable outcomes.

The state of implementation of cotton sector reforms varies widely from country to country. Serious structural reform of cotton sectors in East and Southern Africa (ESA) began in the mid-1990s. Reform in West and Central Africa (WCA) has been slower, for a complex set of reasons related to both domestic and international concerns; among the latter, the case submitted to the World Trade Organization (WTO) by the Cotton-4 countries (Benin, Burkina Faso, Chad, and Mali) regarding market distortions caused by subsidies in Organisation for Economic Co-operation and Development (OECD) countries has given a political dimension to the issues in the sector, and figures prominently among the reasons for resistance to reform in some countries.

Resistance is also due to the perception—genuine or not—that the impact of reforms on sector performance, and especially on small farmers, has been at best mixed. A number of West African leaders and policy makers strongly feel that the reforms are likely to create major social problems, and the results in countries that have implemented reforms, particularly in ESA, do not make a strong case for privatization and liberalization of the cotton sectors. Given the complexities of reform programs and the uncertainties and fluctuations in the world cotton market, it is also difficult to establish clear causal links between structural changes, risks faced by cotton companies after liberalization, and outcomes at the farm and sector levels.

An abundant literature has been produced in recent years on cotton policies in Africa. However, there has been little comparative analysis of the actual outcomes of cotton sector reforms as measured by growth and poverty reduction, and the main lessons to be drawn to inform future reform processes. Also, only limited attempts have been made to bring together, compare, and assess reform experiences from WCA and ESA. Bridging these knowledge gaps in an effort to better advise governments on cotton sector reform programs and policies provides the fundamental rationale for this study.

A number of SSA cotton sectors, especially in WCA, are currently facing serious short-term financial difficulties. It is important to clarify that the purpose of this report is *not* to provide quick solutions to these short-run problems. Rather, it is to step back, build up a reliable broad assessment of cotton sector performance from detailed empirical information, and thereby provide guidance for the design of strategies that will address the long-term challenges of cotton production and marketing in Africa.

Finally, to avoid even the appearance of ideological bias, the authorship of this study has been entrusted to a diverse and independent team of researchers and experts in the field of cotton. Evidence from the analysis is reported regardless of whether it confirms previous theories and hypotheses. Interpretations are suggested, but never imposed, and some care is taken to identify assumptions that drive the analysis. For this reason, the authors, not the World Bank or other bilateral donors, bear individually the full and final responsibility for the content of this report.

# Acknowledgments

This study was carried out by a World Bank team led by Patrick Labaste (Lead Agricultural Economist, Sustainable Development Department, Africa Region [AFTSD], World Bank). The main authors of the report are David Tschirley (Michigan State University) and Colin Poulton (School of Oriental and African Studies, University of London). Other authors are Nicolas Gergely (consultant), John Baffes (Development Economics Vice Presidency, World Bank), Duncan Boughton (Michigan State University), and Gérald Estur (consultant, marketing and quality). The draft report was edited by Julie Dana (Agriculture and Rural Development/Commodity Risk Management Group [ARD/CRMG] World Bank). Review and editing of the final report were performed by Patrick Labaste, Colin Poulton, and David Tschirley. Julie Dana and Christophe Ravry (AFTSD, World Bank) provided inputs at various stages of the study.

Preliminary desk reviews of country cases were carried out in September and October 2006 and discussed at a workshop held in Washington in November 2006. The second phase of the study took place between January and June 2007 and entailed field visits in most of the countries in the sample. A second workshop was held in Washington in April 2007 to share the findings of the field work, country case study analysis, and emerging findings from the comparative analysis as a basis for the preparation of the study's report.

This work was funded by the World Bank and by contributions from bilateral and multilateral trust funds, particularly from Belgium (Belgian Poverty Reduction Program), the Netherlands (Bank-Netherland Partnership/CRMG), the Swiss Secretariat for Economic Affairs (CRMG), and the European Union (All ACP Agricultural Commodities Programme). The team is thankful to the peer reviewers—B. Losch (AFTSD), A. Touré (AFTSD), C. Delgado (ARD), S. Jaffee (ARD), and B. Shamsiev (ECSSD, World Bank)—as well as to Stephen Mink (AFTSD) and Quentin Wodon (Poverty Reduction and Economic Management Group, World Bank), for their useful advice and guidance.

The authors want to thank the various individuals in the cotton business, in Sub-Saharan Africa and elsewhere, who provided advice and information for this study. It is not possible to give a full list of contributors, but the team would like to particularly acknowledge the inputs and contributions provided by R. Chaudhry and A. Gruère (International Cotton Advisory Committee); M. Cour, M. Le Grix, and L. Humbert (Agence Française de Développement); P. Texier (DAGRIS); A. Gruson (SODECOTON); D. Babin, C. Konaté, and I. Coulibaly (CIRAD/CMDT); D. Dakou (SOFITEX); L. Godart (SOCOMA); W. Maro, J. Kabissa, and M. Mtunga (Tanzania); B. Hanyani-Mlambo and J. Battershell (Zimbabwe); S. Kabwe and T. Isherwood (Dunavant/Zambia); and the management of Cargill/Zambia.

Many thanks to colleagues in the World Bank field offices: C. Ngomba, Y. Sangho, H. Gordon, and I. Nébié who facilitated the field visits in Cameroon, Mali, Tanzania, and Burkina Faso, respectively, as well as to N. Ahouissoussi and A. Mwakanawele, who provided comments on country case studies, and M. Sadler who commented on the final draft report. The team also expresses its appreciation to A. Diawara, M. Barton-Dock, K. Brooks, and J. McIntire from the World Bank who supported the initiative right from the beginning and made this undertaking feasible.

The team is also grateful to W. Olthof (Directorate General for Development, European Commission) and V. Gnassounou of the Secretariat of the African, Caribbean and Pacific Group of States, in Brussels, for their interest and support in this research project.

Finally, a special word of thanks to those who provided administrative support throughout the process and contributed to the creation of the report: V. Vaselopulos, H. Page, M.-C. Fundi, M. Diallo, and P. Sadé of the World Bank; and J. Keel of Michigan State University.

# Abbreviations

| | |
|---|---|
| ¢ | US cent |
| AMA | Agricultural Marketing Authority |
| ASARECA | Association for Strengthening Agricultural Research in Eastern and Central Africa |
| CDF | Cotton Development Fund |
| CDO | Cotton Development Organization |
| CDT | Cotton Development Trust |
| CFA | Communauté française d'Afrique (French community of Africa) |
| CFAF | CFA franc |
| CFDT | Compagnie Française pour le Développement des Fibres Textiles |
| CIF | Cost, Insurance, and Freight |
| CIRAD | Centre International de Recherche Agronomique pour le Développement |
| CMB | Cotton Marketing Board |
| CMDT | Compagnie Malienne de Développement des Fibres Textiles |
| CNA | Companhia Nacional de Algodão (Mozambique) |
| CORAF/WECARD | Conseil Ouest et Centre Africain pour la Recherche et le Développement Agricoles |
| DAGRIS | Développement des Agro-Industries du Sud |
| € | euro |
| ESA | East and Southern Africa |
| EU | European Union |
| FAO | Food and Agriculture Organization of the United Nations |
| FF | French franc |
| FOB | Free on Board |
| FOT | Free on Truck |
| GDP | gross domestic product |
| GM | genetically modified |
| GTZ | German Agency for Technical Cooperation |
| ha | hectare |
| HUICOMA | Huileries Cotonnières du Mali |
| ICAC | International Cotton Advisory Council |
| IMF | International Monetary Fund |
| IPC | inter-professional committee |
| IRCT | Institut de Recherches sur le Coton et les Textiles. |
| kg | kilogram |
| mm | millimeter |
| mt | metric ton |
| OECD | Organisation for Economic Co-operation and Development |
| SN SITEC | Société Nouvelle SITEC |
| SSA | Sub-Saharan Africa |

| | |
|---|---|
| T sh | Tanzanian shilling |
| US | United States |
| US$ | U.S. dollar |
| U sh | Ugandan shilling |
| WCA | West and Central Africa |
| WTO | World Trade Organization |

# Executive Summary

Cotton is a rare economic success story in Sub-Saharan Africa (SSA). While the continent's share of total world agricultural trade fell by about half from 1980 to 2005, its share of world cotton exports more than doubled. The crop is a major source of foreign exchange earnings in more than 15 countries of the continent and is a crucial source of cash income for millions of smallholder farmers and their families. In some countries, especially in the Sahel, there is no short- to medium-term cash crop substitute for cotton for small farmers.

Throughout SSA, cotton sectors face major challenges pertaining to competitiveness and sustainability. In parts of West and Central Africa (WCA) sectors are experiencing financial crises brought on by years of declining productivity throughout the sectors, compounded by unfavorable external factors (exchange rate fluctuations and market distortions). Because of the size of the sectors—a function of past success—these crises pose serious threats to the rural economies and macroeconomic stability of the countries. Problems facing sectors in East and Southern Africa (ESA) do not have the same macroeconomic ramifications. Nevertheless, some of the best performers in that region are currently going through credit default crises, and all face important challenges to creating a solid basis for sustained growth over time.

Since the early 1990s, governments of most cotton producing countries in SSA have been implementing sectoral reforms, often with the support of the World Bank and other development institutions. These reform processes generally entailed disengaging the state, facilitating greater involvement of the private sector and producer organizations, ensuring greater competition in input and output markets, improving productivity through research and development and technology dissemination, and seeking value addition through market development and processing of cotton lint and by-products.

The pace and trajectory of cotton reform programs have varied widely from country to country. Profound structural reforms were initiated in the early to mid-1990s with the privatization and liberalization of the cooperative-based systems in Tanzania and Uganda, and the elimination of single channel systems in Zambia and Zimbabwe. The first two have continued to see periodic structural reforms as they had to deal with the problem of input credit provision to smallholder farmers in sectors with many ginners; Zambia and Zimbabwe have seen less radical policy change but have struggled with the problems caused by new entrants. Reform in WCA has been more recent and slower paced, as the sheer size of the sectors and the greater role of the state has made reforms more difficult, both from political and practical perspectives. In several countries where the reform process is less advanced there is a common perception among policy makers and many stakeholders that the experiences of reforms elsewhere, especially in ESA, have resulted in unsatisfactory outcomes and/or patterns of near-term disruption.

*This study was therefore motivated by the perceived need for a broad, empirical, analytically based assessment of reform experience across a range of African countries that would yield lessons and guidance to policy-makers, other local stakeholders, and interested donors agencies.*

The conviction that such an effort was worthwhile emerged from four observations:

- First, much of the public debate on cotton in recent years has been excessively focused on particular, highly visible, and sensitive issues, such as OECD and Chinese subsidies to their cotton sectors While certainly important, such subsidies are not the prime determinants of long-term competitiveness of cotton production and trade in SSA.

- Second, the policy dialogue on the serious challenges facing African cotton has often been highly polarized. In the Francophone West Africa, for example, sector stakeholders have typically focused on the need to preserve input credit and extension systems, while donors have focused primarily on cost efficiency and long-term sustainability. In fact, it is clear that both issues are equally important and need to be considered together.

- Third, relatively little sustained attention has been paid to the precise institutional structure that post-reform sectors could take or why a particular structure might be preferable. Yet any proposal for change must address such issues if reform is to have a reasonable chance of achieving desired goals.

- Finally, very little has been done to comparatively assess the differing experiences of WCA and ESA and draw lessons across the regions.

## Methodology and Conceptual Framework: A Typology of African Cotton Sectors

This comparative study was undertaken to examine the complex issues raised above, with a view to bringing fresh insights to inform and guide decisions rather than fueling old controversies To implement it study, the World Bank brought together a large and diverse team of experts and researchers with extensive experience in cotton sectors across the continent. The review was based on detailed case studies in nine of the main cotton producing countries of Sub-Saharan Africa: Benin, Burkina Faso, Cameroon, and Mali in WCA; and Mozambique, Tanzania, Uganda, Zambia, and Zimbabwe in ESA. The study analyzes systematically and thoroughly the relations between sector organization and sector performance and outcomes, with the aim of establishing evidence-based causality links between a given sector organization and a series of performance indicators. On the other hand, it does not pretend to provide detailed prescriptions to guide further reform processes in individual countries. Stakeholders in the various countries will need to draw implications from the comparative analysis and apply pertinent lessons to their local circumstances.

The study followed a four-step process. It first drew on available experience to develop a conceptual framework that would generate testable hypotheses. Next, it developed an analytical framework that included the characterization of clearly distinct sector types—based on the market structure and associated framework for seed cotton—and a set of empirical performance indicators with which to test the hypotheses. Third, the team used the country case studies, together with a broad range of information sources (literature review, historical analysis, key informant interviews, focus group interviews, secondary quantitative data, and newly collected data) to "tell the story" of reform in each country and inform the comparative analysis. This entailed computing the indicators; benchmarking performance, past and current, by country and sector type; and drawing conclusions on the major drivers of that performance, particularly with regard to sector structure.

*The conceptual framework for this comparative analysis of cotton sectors rests on the idea that economic systems benefit from both competition and coordination, but that in the real world of imperfect markets and weak states there is likely to be a trade-off between them.* Conventionally produced cotton is an input-intensive crop, so Africa's frequently failing input and credit markets present a particular challenge to its development. National quality reputation for cotton also remain important, providing a second major justification for sector coordination. Competition is important to ensuring efficiency and equitable sharing of benefits between buyers and sellers. Yet too much competition will make it difficult or impossible for stakeholders to engage in the coordination needed to provide important services such as quality control, input credit, research, and extension. A well-functioning cotton sector is one that strikes a balance between these competing needs, providing sufficient benefits to all stakeholders so that the system is able to maintain itself and grow.

This hypothesized trade-off is at the center of the organizing feature of this book: *a typology of cotton sectors in SSA based on the structure of the market for the purchase of seed cotton and of the regulatory framework in which farms and firms operate.* Market structure can be defined by the nature of players and entities in the sector, together with the distribution of roles and power between them. The regulatory framework is the set of rules, regulations and other legal instruments that are imposed on participants in the sector to enable it to operate and limit conflicts. These two pillars of the framework (market structure and regulatory framework) are based on the observation that structure has a fundamental influence on the balance that a country is able to strike between competition and coordination, and that regulatory frameworks can modify the competition-coordination trade-off but must deal with structure as it exists. Together with basic characteristics common to nearly all SSA economies (such as the predominance of undercapitalized smallholder farmers, widespread failure of input and credit markets, and weak legal systems), these two factors establish the opportunities and constraints that cotton sector stakeholders operate within.

Figure 1 below shows the five sector types identified in the typology, along with each country's current location and its evolution over the recent past. Solid arrows depict changes that have taken place in sector organization since the mid-1990s; dashed arrows suggest changes that may be underway. The lines showing recent changes in sector organization illustrate the point that reform is not a one-off event and that most sectors are still seeking an optimal structure to cope with the challenges of remaining internationally competitive.

**Figure 1 Location of Cotton Sectors within African Cotton Sector Typology**

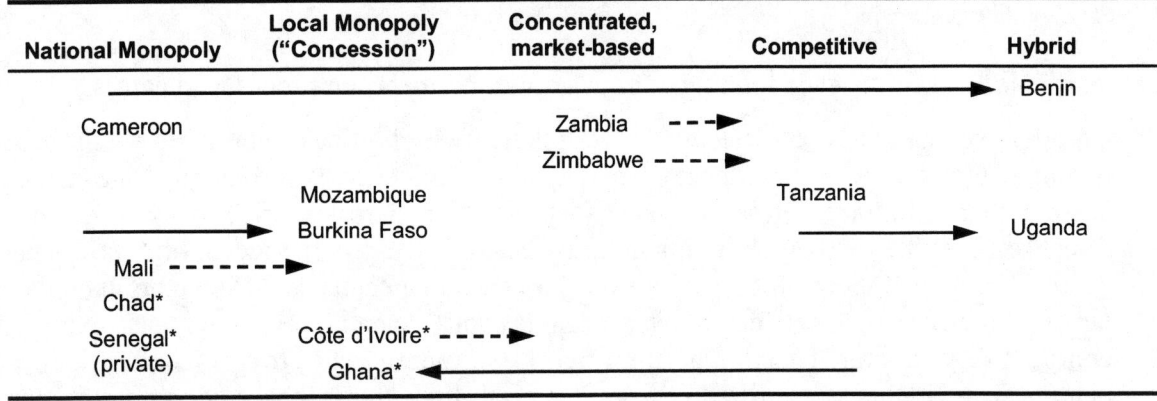

*Note:* * Not included in this study

National monopolies are now found only in WCA and, with the exception of Senegal, are owned and operated by public or mixed companies. Local monopolies have more than one firm, but establish geographical areas within which each has an exclusive right to purchase all cotton (and, typically, a responsibility to promote it). Ginning companies compete directly with each other in concentrated and competitive sectors. The basic difference between these systems lies in the number of ginning firms. Thus, Zambia and Zimbabwe were both, until fairly recently, effectively duopsonies in which the top two firms accounted for 90 percent or more of seed cotton purchases. By contrast, in Tanzania there are about 30 ginners, the top 5 of which have only about a 40 percent market share (and these top 5 typically change from year to year).

*This difference in market concentration is also associated with an important difference in the nature of competition across the two sector types.* In concentrated sectors, firms compete (on reputation) for the right to transact with producers through the coming season. By contrast, in competitive sectors there are few incentives for preharvest service provision. Instead, they compete for seed cotton on the basis of price at harvest time. Finally, hybrid systems are a potentially diverse group, emerging either out of attempts to liberalize a national monopoly (Benin) or to solve the problems unleashed by liberalization in a sector with a competitive structure (Uganda).

Performance across the different sectors and sector-types was assessed through a range of efficiency, effectiveness, and distributional indicators, some at the micro (i.e. farm or enterprise) level and others at more sectoral or macro levels. Some of these indicators are readily quantifiable, while others are based on qualitative 'order of magnitude' judgements. They include process indicators (prices paid to farmers, services delivered to them, quality management, technology creation and dissemination, valorization of by-products), and outcome indicatorsfor stakeholders (revenue and profitability at farm level, economic efficiency, value added, macro impacts), as well as an assessment of the logical links between these. The conceptual framework generates clear expectations about how different sector types would perform on many, though not all, of these indicators. It also recognizes that structural factors may not fully determine outcomes, as other factors will also influence competitiveness, incentives, and the acceptability and impacts of reform measures.

## Empirical Results: Strong Links between Sector Structure and Performance

*In large measure, the analysis showed that sector structure has a major and predictable influence on performance.*

This is particularly the case when looking at how a sector performs on process indicators.

- Competitive, market-based systems deliver relatively high prices to farmers, but are weak on input credit provision, extension, and quality. Evidence is strong in Tanzania and Uganda that, within market-based systems, competition increases prices paid to farmers, a direct result of intense competition among companies. Despite the recent high prices paid by the WCA monopoly systems, taking a 20-year perspective, WCA sectors have been largely outperformed on price by Tanzania and Uganda. However, as expected, competitively structured sectors perform poorly on input credit provision, extension, and quality. This finding stems directly from the great difficulty in coordinating across more than a few companies, whether this coordination is to prevent side-selling or to agree on discounts to be paid for poor quality seed cotton., *on how a sector performs on its process indicators*:

- **Concentrated and monopoly (national or local) sectors *can* perform well on prices paid to farmers,** but such performance depends on the strategic priorities of dominant companies (which can vary over time), on the existence of political interference (if any), and on the voice of cotton farmers in price negotiations. Since 2000, concentrated sectors (Zambia and Zimbabwe) have performed relatively poorly on prices to farmers, while national monopolies have paid unsustainably high (yet politically-backed) prices that have been an important contributor to these sectors' fiscal crises.

- **Concentrated sectors do well on quality management and, to a certain extent, delivery of services to farmers**. Concentrated sectors perform best on quality. They also provide input credit and extension advice to large shares of cotton farmers, although farmer coverage is not as complete as in national and local monopolies.

- **National and local monopolies in WCA have been able to provide input credit and extension to a large number of farmers** and achieve relatively high yields as well as high and fairly stable credit repayment rates. However, there are reasons to believe that the quality of extension assistance in these monopoly systems has probably declined over the past twenty years, while farm yields have been stagnant or declining since the mid-1980s.

- **Research efficacy is not clearly linked to sector type**. One hypothesis was that a small number of large companies should be better placed to influence research performance than multiple small ginners, but in practice this was not borne out. The main reason for this is that governments have been slow to allow private ginners to contribute to research management, even where they have been allowed to assume responsibility for most other aspects of national cotton production and marketing; most research programs remain firmly within the public sector.

- **Low valorization of cotton seeds can be observed in monopoly systems**. In the case of by-product valorization, WCA sectors (especially Mali and Burkina Faso) receive low prices for cottonseed, an outcome related to the history of vertical integration within the sector. Otherwise, major determinants of prices received include whether a country is landlocked and the strength of local demand for cake for cattle feed. In any case, by-product valorization has so far received insufficient attention in the study of cotton sectors and is an area where value could be added and (re)distributed to farmers through greater opening of the sector and competition.

Performance on outcome indicators is less clearly tied to sector type, because many outcomes are a function of more than one process, and a given sector type might perform well on some processes but poorly on others. Returns to farmers are a clear example:

- Competitively structured sectors pay high prices to farmers but are unable to provide input credit or extension; as a result, they tend to generate low yields. They also score poorly on lint quality, which limits the price advantage they can pass to farmers.

- Concentrated sectors do better on input credit and extension (and thus on yields), and also do well on quality. But they may pass little if any of the quality premium on to farmers, and may have a tendency to charge higher than market rates for the inputs they provide. As a result, returns to farmers are similar in Zambia's concentrated system and Tanzania's competitively structured one.

- Sectors performing best on returns to farmers are those that have benefited from many years of sustained investment in research and extension, and that therefore have been able to raise the productivity of large numbers of farmers; most WCA countries and Zimbabwe in ESA fall into this category. However, in WCA, this performance has declined during the last two decades, as have the returns for a majority of farmers.

During the present decade, national monopolies have performed very well on one macro indicator (valued added per capita), but very poorly on another (net budgetary contribution per capita). The positive performance on per capita value added in Burkina Faso and Mali has come at a steep cost to the rest of the economy, especially to the state budget, particularly in recent years. Tanzania's competitively structured sector performs very well on value added per capita during some years, but poorly during others, as a result of highly variable production. A key insight from the work is that this variable production is a direct result of Tanzania's sectoral structure: because farmers do not have the incentive of in-kind input credit to produce cotton, they move in and out of the crop based on expected price, much like they would with any other cash crop.

Company efficiency is one outcome indicator clearly tied to sector structure: companies in market-based sectors are most efficient, with competitively structured sectors more efficient than concentrated sectors, but with both greatly outperforming companies in local and national monopolies. The combination of high farmer prices during the last six to seven years plus relatively inefficient companies—that is, those with high operating costs—means that the WCA monopolies are, by a substantial margin, currently the least "internationally competitive" sectors in the study. Free-on-truck cost to value ratios in WCA monopolies range from 0.98 to 1.15, compared with a range of 0.76 to 0.88 in all other countries except Uganda. While it is clear that the WCA national monopoly model has generated strong returns to very large numbers of farmers, poor incentives for cost efficiency have undermined the international competitiveness of these sectors as well as their contribution to the wider economy.

In Zambia and Zimbabwe, and also in Mozambique, the "competitive" cost-to-value ratios have been achieved in part because of the low prices paid to farmers. In Mozambique (which scores 0.79 by the study's measure) the seed cotton price has been 20 percent to 30 percent lower than prices in all other ESA countries. In Zambia (the most internationally competitive sector in the study at 0.76), the seed cotton price to farmers is substantially higher than in Mozambique, and not far below that in Tanzania in absolute terms, but reflects little of the substantial quality premium that Zambian companies receive on the international market.

## Core Challenges: Performance of African Cotton Sectors in the Global Context

The comparison of the nine cotton sectors of the study sample concluded that no single market sector type performed so well that it can be considered the benchmark for all others. There are strong correlations between sector type and performance results, yet there is no ideal model among the study countries and trade-offs have to be considered. There are clear indications that factors in addition to sector structure do have an influence. A given sectoral type can perform well or poorly on final outcome indicators, and this performance is strongly influenced by history (including past investments), culture, managerial effectiveness (which is partly a function of culture and individual personalities), and agro-ecological endowments. Nevertheless, sector type (market structure and associated regulatory framework) does say a great deal about the key challenges that will be most difficult for a sector to meet, and about the most promising approaches for dealing with those challenges. For example, input credit, extension, and quality will be problems in competitive systems; prices to farmers will tend to be low in concentrated sectors; company efficiency will tend to be poor in monopolies.

The high intrinsic quality of African fiber, the fact that it is handpicked, and the low unit production costs of its smallholder production base give African cotton important growth potential on the world market in the long run. *However, an assessment of relative performance among African countries and what they need to do to be competitive reveals that all sectors show productivity and performance gaps on a global scale and therefore generally lag well behind the best performers in the world.* Across the full spectrum of African cotton industries, core challenges for competitiveness and profitability have emerged. All African cotton sectors face increasing competition from other countries and continents and from synthetic fibers, and thus face continual pressure on prices. African cotton sectors must therefore continually strive to cut costs, raise productivity, improve lint quality, and add value if they are to maintain attractive returns to farmers and to make a positive contribution to national poverty reduction goals. To achieve this, cotton sectors need to improve their performance on critical factors such as the responsiveness and efficiency of research and extension; technology transfer in areas such as genetically modified strain dissemination; lint quality management and marketing; soil conservation; and technical support to farmers and farmer organizations.

Effective strategies for African cotton sectors should therefore combine institutional innovations and reforms with necessary investments in key public goods. The three broad objectives that all African cotton sectors should pursue are (1) achieving greater value through improved quality, marketing, and valorization of by-products, (2) bridging competitiveness gaps through farm-level productivity and ginning efficiency, and (3) improving sustainability through institutional development and capacity-building of stakeholders, as well as strengthening of governance and regulatory structures and management systems. Some of these actions could usefully be tackled at a regional level, as well as nationally, and could benefit from donor support.

The analysis also showed that a country's history, current sector type and political imperatives also have a significant influence on the feasible path of institutional change over time. For example, concentrated (market-based) sectors emerge from the analysis as perhaps the best performers, doing well on a wide range of indicators. Yet in the current setting in SSA, marked by institutional and human capacity weakness, these sectors have a difficult time maintaining their concentrated structure. They tend to slide toward greater competition, with predictable declines in performance on input credit, extension, and quality. A worrisome finding is that, as competition increases in concentrated sectors, input supply and quality control may decline *before* prices paid to farmers show noticeable signs of improvement. Such a pattern is of special concern because these negative impacts are quite difficult to reverse.

## Structure as a Key Factor of Competitiveness and Sustainability of African Cotton Sectors

Implicit in reform programs in cotton and other sectors in Africa to date has been the notion that structure matters, at least insofar as it promotes or impedes competition. One of the main conclusions of this analysis is that structure does explain a good share of the variability in sector performance. This is an extremely important finding for decision makers, as well as a clear encouragement for governments to pursue reforms of their cotton sectors as a means of ensuring future competitiveness. At the same time, this analysis recognizes that other factors, such as history, managerial competence, geography, politics, macroeconomics, developments in competing or complementary sectors, and even luck, also play a role that cannot be ignored. It means that structure does not uniquely explain everything either, and that there are some common core challenges facing all African cotton sectors.

*The ultimate objective of reform is to strengthen the competitiveness of cotton production in the context of the global world market and thereby to ensure long-term, sustainable, and equitable growth for these major sectors of many African economies.* The broad policy objectives include farmer welfare, industry innovation, technical and economic efficiency, and value addition. However, there may be some trade-offs among these due to the tension between competition (incentives) and coordination (controls) within the different systems. The comparative analysis based on the typology showed the strengths and weaknesses of the various sector types. However, this does not mean that the pros and cons of each model are all of equal nature and magnitude thereby tending to offset each other. This would be a very static interpretation of the necessary trade-offs between coordination and competition. To the contrary, the review highlighted important differences in long-term viability across sector types:

- At one extreme, most single-channel systems have become financially unsustainable, not only because of the current exchange rate of the CFA franc, but mainly because of their inability to adapt, manage, and innovate quickly in response to changing circumstances. At the other extreme, competitive models have serious drawbacks in the current SSA context—that is, weak markets for inputs and financial services—which means that they are environmentally unsustainable from an environmental perspective.

- Local monopolies and concentrated systems may offer better prospects for the future, but subject to a number of critical conditions such as the careful selection of investors with long-term commitments to the sector's development, the setting up and enforcement of regulatory frameworks adapted to the sector's needs, the strengthening of input and credit markets, and the building up of inter-professional organizations to ensure broadly based representation in sector management.

## Insights on Major Challenges and Feasible Reform Paths for Specific Sector Types

Cotton sectors are facing broad generic types of challenges---value added, competitiveness, sustainability---and the need to (re-)arrange pertinent rules and incentives to meet these challenges. Reform programs should therefore be designed to help specific sectors face their own peculiar challenges. In effect, relatively little sustained attention has been paid so far to the precise structure that post-reform sectors could take or why a particular structure might be preferable.

The report concludes that problems of cost inefficiency and managerial dysfunction in *national monopolies* are serious enough, and the politics of improving performance under such monopolies is complicated enough, that most of them need to move toward a different sectoral structure. But this conclusion should not be applied dogmatically. Cameroon, with relatively good performance to date, may be able to maintain its national monopoly to good effect, as long as it reforms its price-setting process and continues to promote productivity.

If national monopolies do choose to change, the direction of change is likely to be toward local monopolies or concentrated, market-based systems. The report concludes that moving in WCA to fully privatized markets allowing competition among companies, even if the market is initially very concentrated, is risky because of the possible instability of concentrated sectors. If, instead, these sectors can use the local monopoly approach to develop sound regulatory mechanisms and build the operational capacity of farmer organizations, concentrated and eventually competitive systems could perform well. During the local monopoly phase, care must be taken to ensure that private companies play a greater role in price setting and other decision making than they have so far played in Burkina Faso; price setting must also be done in a way that provides reasonable assurance to companies that, if they operate efficiently according to international standards, they will be able to earn a reasonable return on their investment. Clear rules for evaluating and re-tendering concession areas are also needed, as this has been a clear failure in Mozambique.

The key challenge in Mozambique's *local monopoly* sector is how to create incentives for good company performance. In the absence of strong farmer associations, these incentives have to come from improved rules governing tendering and re-tendering of concessions, procedures for monitoring performance of concessionaires, and careful selection of companies.

The key challenge for *concentrated sectors* is to develop a regulatory approach with three characteristics:

- Clear and transparent licensing rules must be developed that strictly specify the capabilities and conduct of firms wishing to participate in the sector, to defend the ability of firms to coordinate on input supply, extension, and quality control.

- These rules must be strictly enforced, but must allow enough prospect of entry that incumbents have an incentive to maintain attractive seed cotton prices.

- Given the problems of relying entirely on the threat of entry to discipline incumbent firms, it may also be desirable to develop price-setting mechanisms that are more formalized than the price leadership that has prevailed in concentrated systems so far.

***Competitive sectors*** perform poorly on coordination for service provision. The book concludes that such coordination as does occur must come from a central body and that the state needs to play a key role within this body. This is in contrast to local monopoly or concentrated systems, where inter-professional committees dominated by ginners and farmers have more potential to adequately manage the sector. Given the well-known problems of such state involvement, the accountability of regulatory bodies toward ginners and farmers in competitive systems needs to be strengthened. Because incentives are very limited within competitive sectors for individual ginners to support long-term programs for productivity growth, the state and the ginners' association may have to work with other actors (local government or donors, for instance) to develop programs that enhance the asset base of farmers and also generate benefits beyond the cotton sector. The review suggests that if competitive sectors (as in Uganda) move to hybrid structures, such approaches need to avoid protecting ginners entirely from competitive pressure; this conclusion is based on Uganda's experience, where entrenched overcapacity eliminates cost advantages the sector would otherwise have. Tanzania's agro-ecological and population settlement characteristics have so far protected it from the need to take the type of radical measures that Uganda has experimented with for input credit provision. However, if yields begin to fall as a result of declining soil fertility (or possibly one day to increasing pest pressure), and if it wants to realize its potential more fully, the country may need to consider moving to a more coordinated approach.

Finally, when referring to the challenges faced by the various sector types, it is important to recognize that governments, private companies, farmers, and service providers may experience these challenges in very different ways. One of the next steps in the reform processes is therefore to gain consensus on what these particular challenges are for specific actors and what further investment is needed to address them.

# SECTION I
# INTRODUCTION AND MARKET CONTEXT

# Chapter 1: Introduction
## *David Tschirley*

Cotton in Sub-Saharan Africa (SSA) is at once a major success story and, in much of the continent, a focus of intense concern. While the continent's share of world agricultural trade fell by about half from 1980 to 2005, its share of cotton trade more than doubled.[1] Production grew three times more rapidly in SSA over the period than it did in the rest of the world. Cotton is predominantly a smallholder crop, with over 2 million poor rural households in SSA depending on it as their main source of cash income. Among export crops with substantial smallholder farmer involvement in SSA, cotton ranks second in value only to cocoa, and cotton's production is spread more widely across the continent. The profitability of cotton production and processing in Africa has large and widespread impacts on rural growth and poverty in the continent, and as a result, the challenges faced by the sector are serious.

Unusual for African export crops, cotton is also produced in several countries of the developed world and in China. Large subsidies to cotton farmers in many of these countries, combined with the obvious role that cotton plays in the livelihoods of millions of poor African farmers, has helped make the crop a major issue in world trade negotiations. The Overseas Development Institute (ODI 2004) shows that subsidies to cotton farmers in the United States during 2001/02 were equivalent to about 50 percent of the world price; in China and the European Union, these figures were about 25 percent and 100 percent, respectively. The total value of subsidies is highest in the United States, where about 25,000 cotton farmers received an average of about US$2 billion per year between 2001 and 2003, equal to about 60 percent of the national GDP of both Burkina Faso and Mali.

Formal complaints under the World Trade Organization (WTO) about these subsidies began in 2003 with Brazil, which challenged US subsidies and won its case in 2004. Also in 2003, Burkina Faso presented the WTO with a cotton proposal on behalf of itself, Benin, Chad, and Mali calling for the eventual elimination of all developed-country cotton subsidies, coupled with financial compensation for cotton farmers in developing countries. Within Africa, public debate about subsidies has focused almost entirely on these four West African countries, know as the "C4" (Cotton-4).

Predating this trade and subsidies debate has been another debate, now lasting more than two decades, on whether the highly integrated approach to cotton supply chain development in countries of West and Central Africa (WCA) needed to be reformed. The WCA approach, which typically featured "single-channel" systems built around public monopoly cotton companies, has driven tremendous growth in cotton production in the region; the International Cotton Advisory Council (ICAC) data indicate that total lint production in the CFA Franc Zone **rose** from 50,000 tons in 1960 to about 220,000 tons in 1980, to an average of about 1.1 million tons in 2004 and 2005. The crop has also played a major role in rural development, facilitating input supply for other crops in cotton zones and helping farmers invest in animal traction and other equipment that improved overall farm productivity and incomes.

However, these single-channel systems have also suffered from serious and perhaps growing problems. During the years immediately following the devaluation of the CFA franc in 1994

(Pursell 1999; Badiane et al. 2002), cotton sectors in WCA were seen to pay lower prices to farmers than sectors with more competitive arrangements. Many studies commented on the stagnation of WCA farm yields starting about 1990, although these yields remained higher than in most other areas of Africa. The parastatal ginning companies were also seen to be increasingly inefficient and opaque in their operations (Pursell 1999; Badiane et al. 2002; World Bank 2007). In Mali, farmers boycotted the crop in 2000/01 because of low prices and perceived corruption within Compagnie Malienne de Développement des Fibres Textiles (CMDT), and top managers in CMDT were eventually sent to jail for financial mismanagement.

In a world market where real prices have fallen by about half since 1980,[2] the problems described above can threaten the survival of cotton production, processing, and trade. These concerns have come acutely to the fore since the beginning of the 2000s, as high prices to farmers, combined with high operating costs of the ginning companies and stagnant farm yields, have led to massive sectoral deficits in most countries. In Burkina Faso and perhaps other countries, these deficits threaten nationwide macroeconomic stability. Meanwhile, farmers continue to complain that the prices they receive are too low.

The debate about how to deal with these problems is rooted in several factors at the intersection of characteristics of cotton as a crop and the rural setting in much of SSA. First is the widely appreciated fact that cotton production requires substantial use of external input, specifically treated seed, fertilizers, and insecticides.[3] A second factor is that markets in SSA for input, especially credit for input, frequently fail for smallholder farmers. While seed and fertilizer for a crop like maize may be relatively available in markets and frequently purchased by smallholder farmers, specialized insecticides and seed treatments for cotton are less likely to be available, and credit is almost never accessible by unorganized smallholders. Additionally, because cotton is produced in a highly competitive export market, efficiency is paramount throughout the chain. At the farm level, farmers must use the right input in the right way if they are to earn reasonable returns from the crop, and if they are to produce enough product to sustain the ginning companies. Control over the input mix and extension assistance to ensure proper use are issues in which ginners have vested interests. Bundling input and extension into a package creates efficiencies for the distributor and as a result, most approaches to the input credit problem have featured interlocked transactions. A ginning firm wishing to purchase the farm output provides some level of extension advice along with input to farmers on credit, and attempts to recover the credit upon purchase of the farm's product.

Such arrangements, known as contract farming or out grower schemes, have governed production of a wide range of cash crops throughout the developing world for many decades.[4] When effective, these arrangements allow smallholder farmers to profit from a crop they might ordinarily not be able to plant and allow processors to benefit from low costs of production.[5] Yet the conditions under which interlocked transactions can be expected to emerge and persist are relatively restrictive (Delgado 1999; Benfica, Tschirley, and Sambo 2002).[6] Numerous examples exist of failed efforts, primarily related to the inability of processors to recover input credit (Stringfellow 1996; Glover 1990). Though the structure of the cotton market lends itself to contract farming operations, it too has frequently been threatened by acute credit default crises in many countries. Additionally, over the longer term, cotton systems can be undermined by the inability of participants in the supply chain to agree on and develop financing mechanisms for investments in research, extension, risk management, and quality control.

The performance of cotton input credit and extension systems in SSA is strongly influenced by the structure and behavior of the market for seed cotton. Because changes in the structure of the output market are central to any sectoral reform, they have the potential to dramatically affect input-credit-extension systems. It should not be surprising, therefore, that proposed reforms have engendered great concern about possible unanticipated negative effects on these systems. As early as 1988, a major comparative review of cotton sector performance in Anglophone and Francophone countries of SSA concluded that, in West Africa's single-channel systems (which to that time had been far more successful than systems in Anglophone countries), "privatization of input distribution ... should be considered only with the greatest caution, due to the need to link distribution with credit and output marketing" (Lele, Van de Walle, and Gbetiobouo 1989, 31). This review further cautioned about the potential "collapse of the cotton industry in francophone Africa" if research and extension were moved out of existing single-channel systems without viable alternative institutional approaches to ensuring the continuity of these activities. Within Francophone African countries, one important basis for opposition to reform has been fears that input credit and extension would be undermined.

In both the subsidies debate and the debate on structural reform of cotton sectors, little attention has been paid to countries of East and Southern Africa (ESA). Yet production in ESA has been growing steadily, and reached nearly half a million tons of lint in 2004/05. Serious structural reform of cotton sectors in ESA, including the elimination of existing single-channel systems, began in the early 1990s and much has been learned about the process. Reform in countries of WCA has been slower for a number of reasons: the single-channel systems were very strongly established in many WCA countries, stakeholders could point to substantial successes in addition to clear and mounting problems, and the sheer number of farmers involved—and the size of the public companies serving them—made reform difficult from political, social, and commercial perspectives. The developed-world subsidies referred to above also fueled internal resistance to reform. Yet nearly all countries in WCA have made substantial incremental changes in their systems, and some have undertaken (or will soon undertake) structural reforms. To date, few attempts have been made systematically to bring together and assess reform experience from both regions of the continent. This analytical gap, and the potential benefits from such an exercise, provides the fundamental rationale for this study.[7]

This comparative analysis is based on detailed case studies in nine countries of ESA and WCA: Benin, Burkina Faso, Cameroon, and Mali in WCA; and Mozambique, Tanzania, Uganda, Zambia, and Zimbabwe in ESA (figure 1.1). During the period 2004–08, these countries produced an average of 0.98 million tons and accounted for 70 percent of SSA's cotton production (1.35 million tons during that period). In 2005/06, the nine countries in the sample together produced over 1 million tons of cotton lint, the majority of which was exported. This represents 60 percent of total African production and 68 percent of SSA production. The four WCA countries accounted for 70 percent of total production of countries in our sample in 2005/06.[8] Figure 1.2 shows average cotton production for major African cotton producers during 2004–08 (including the nine countries analyzed in this study).

Each case study involved a literature review plus a two-week visit to the country[9] by researchers who already had several years of experience in the sector. In addition to compiling standard information on the historical background, recent changes, and current organization of the sector in each country, the case studies shared three key characteristics that heavily influenced the

3

**Figure 1.1 Map of Africa Highlighting Study Countries of WCA and ESA**

*Source:* Authors.

content of this comparative report. First, each provided a detailed overview of the institutional arrangements in place for key sector services, such as input credit, research, and extension; lint marketing and quality maintenance; and seed cotton pricing. Especially in ESA, the diversity of approaches to these challenges provides great scope for learning. Second, each study developed disaggregated budgets for representative ginners, allowing this analysis to conduct a detailed comparison of the level and structure of ginning costs in each country. Third, each study—except Benin and Cameroon—used a comparable focus group approach to develop detailed farm-level budgets for a range of cotton farmer types. Focusing on the diversity of farm types, behavior, and performance at the farm level provides important insights into cotton's contribution to poverty reduction, into the differential effects of pricing policies, and into the nature and scope of the productivity challenges the sector faces.

**Figure 1.2 Major SSA Cotton Producers (000 tons of lint, 2004-08 average)**

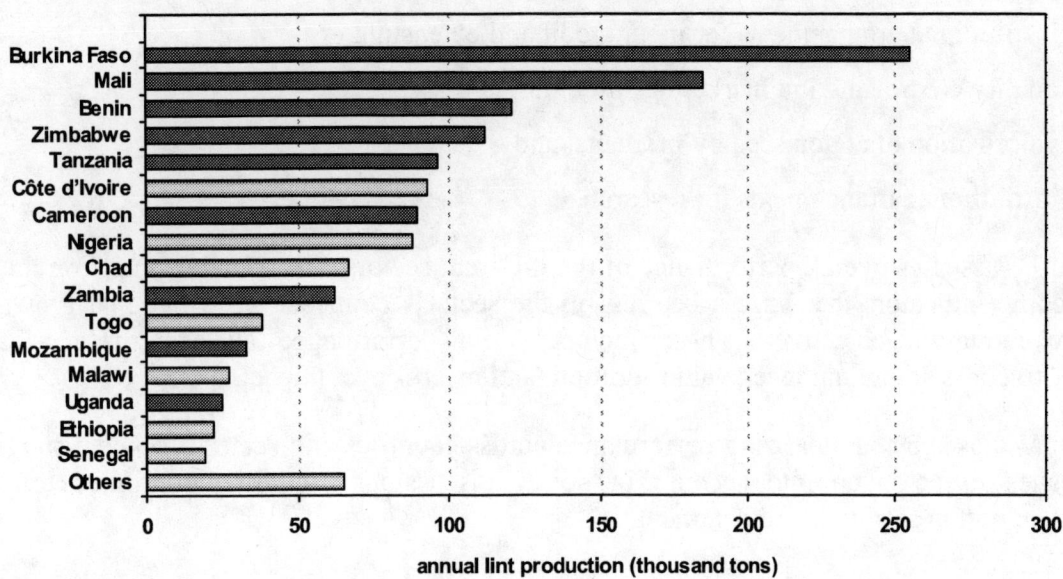

*Source:* ICAC.

*Note:* The list includes countries producing more than 20,000 tons of lint. Red/dark color denotes countries covered in this study.

The term "reform" is widely used but seldom defined in development literature. This book understands reform to be consciously chosen change in the fundamental organization of a sector and related changes in the "rules of the game" under which stakeholders operate. Under this definition, every country in ESA reformed its cotton sector in the early- to mid-1990s, either eliminating single-channel parastatals (Zambia and Zimbabwe) or privatizing cooperative ginners (Uganda and Tanzania).[10] In the sample WCA countries, only Benin and, to a lesser degree, Burkina Faso have reformed their cotton sectors under this definition.[11]

Yet a key theme that emerges from this research is that reform does not presuppose a movement from one stable set of rules of the game to another stable set. The reforms of the 1990s in ESA have been followed by continuous and sometimes dramatic change in every country. Even countries in WCA that have not reformed under this definition have effected substantial incremental change in the ways in which they carry out critical sector activities. An important contribution of this research is showing how the details of institutional design matter, providing a rich sense of the diversity of approaches that have emerged to deal with common challenges, and providing insight into the factors that influence which approach might be chosen under different circumstances.

The report is organized in five sections. The next chapter in section I presents key elements of the current world market setting. Section II provides historical background, outlines the typology of cotton sectors used to form hypotheses about sector performance, and presents a conceptual framework for analyzing sector organization and linking it to outcomes.

Sections III and IV provide a detailed comparative description and performance assessment of nine systems around a set of key themes. Section III focuses on processes, functions, and services that are most directly under the control of cotton companies:

- pricing systems and prices paid to farmers,
- institutional arrangements for input credit and extension,
- quality control and lint marketing strategies,
- valorization of cotton seed by-products, and
- institutional arrangements for research.

Section IV assesses overall performance of the different sectors of the study sample on the basis of outcome indicators that have a bearing on the sector's contribution to national growth and poverty reduction objectives. These include yield performance and returns to farmers, competitiveness at ginning level, value addition, and macro-level impacts.

Section V closes by summarizing performance across countries and sector types and suggesting key issues to be taken into account in policy discussions regarding continued efforts to strengthen cotton sectors in the continent.

# Chapter 2: Market Context
### *John Baffes and Gérald Estur*

African cotton producers are becoming more and more aware that they operate in an increasingly integrated and quickly changing global market. Any analysis of the evolution and performance of cotton sectors in Sub-Saharan Africa (SSA) must therefore be put in this context and must take into account the market dynamics and their implications for the entire supply chain, back to farm level.

This chapter aims to set the stage by providing a strategic overview of the main characteristics of this market. First, the key trends in global supply and demand are presented and discussed. A short analysis follows of price trends and price determinants, particularly with regard to external factors affecting international prices of cotton lint and its by-products, including exchange rate fluctuations and market distortions created by subsidies in some major exporting countries. Specific attention is given to critical factors influencing the demand for cotton lint, such as quality and marketing strategies. The chapter closes with an overview of the market for cottonseed by-products, an increasingly important element of valorization of raw cotton for farmers and for the national economies.

## The Supply Side: Expanding Production and Exports

About three-quarters of cotton is produced by developing countries. Since 1960, world cotton production has grown at an annual rate of almost 2 percent to reach 25 million tons of lint in 2006, up from 10.2 million tons in 1960. Most of this growth came from China and India, whose production quadrupled during that period. Today these two countries account for almost 45 percent of world cotton production and more than half of global consumption. Other countries that significantly increased their production shares during this period were Brazil, Greece, Pakistan, and Turkey (see table 2.1). Some new entrants also contributed to this growth. Australia, for example, which produced only 2,000 tons of cotton in 1960, averaged 0.5 million tons between 1995 and 2005. Francophone Africa produced less than 100,000 tons in the 1960s and now produces 10 times that much. The United States and the Central Asian republics, two of the four dominant cotton producers during the 1960s, have maintained their production at about the same levels, effectively halving their market shares. A number of Central American countries that together accounted for 250,000 tons during the 1970s now produce virtually no cotton at all.

About one-third of cotton production is traded internationally. The three dominant exporters—the United States, Central Asia, and Francophone Africa—account for more than two-thirds of global exports. Overall, Sub-Saharan Africa (SSA) increased its share in world cotton trade from 7 percent in 1960 to 15 percent in 2006. However, the export performances of West and Central Africa (WCA) and East and Southern Africa (ESA) differ considerably: In 1960, WCA accounted for a little more than 1 percent of global exports while today it accounts for more than 11 percent. Exports from ESA, however, have declined from 6 percent in 1960 to 4 percent today.

World prices for cotton have been declining partly as a result of competition on the supply side, which has driven down production costs. Reductions in production costs have been associated primarily with technological improvements resulting in yield increases from 300 kilograms of lint per hectare in the early 1960s to about 700 kilograms of lint per hectare in 2005 (world average). This yield increase reflects the introduction of improved varieties and increased use of irrigation and chemical fertilizers. The spread of genetically modified (GM) seed technology in developing countries and precision farming in developed countries is expected to reduce the costs of production even further.

More than one-quarter of the area allocated to cotton is currently planted using GM varieties, accounting for almost 40 percent of world production. GM cotton in the United States—where it was first introduced in 1996—currently accounts for about 80 percent of the US area allocated to cotton. Other major GM cotton producers are Argentina (70 percent of its cotton area), Australia (80 percent), China (60 percent), Colombia (35 percent), India (10 percent), Mexico (40 percent), and South Africa (90 percent).[12] Countries at a trial stage include Brazil, Burkina Faso (the only SSA country), Israel, Pakistan, and Turkey (*Cotton Outlook* 2005).

Although the last decade has witnessed the expansion of the organic movement in other commodities, cotton has not enjoyed much success. Organic cotton production was introduced in the United States in 1990, when 330 tons were produced. Following a peak of 7,425 tons in 1995, the United States now produces less than 2,000 tons. Currently, the world's two major organic cotton producers are India and Turkey, which together account for two-thirds of global organic cotton production. In 2004, this production was 25,400 tons, only about 0.1 percent of world cotton production.

## Table 2.1 Global Balance of the Cotton Market
*(thousand metric tons)*

| Country or region | 1960 | 1970 | 1980 | 1990 | 2000 | 2005 | 2006 | 2007 |
|---|---|---|---|---|---|---|---|---|
| **PRODUCTION** | | | | | | | | |
| China | 1,372 | 1,995 | 2,707 | 4,508 | 4,417 | 5,714 | 6,729 | 8,078 |
| India | 1,012 | 909 | 1,322 | 1,989 | 2,380 | 4,148 | 4,590 | 5,355 |
| United States | 3,147 | 2,219 | 2,422 | 3,376 | 3,818 | 5,201 | 4,731 | 4,182 |
| Pakistan | 306 | 543 | 714 | 1,638 | 1,816 | 2,089 | 2,115 | 1,845 |
| Central Asia | 1,491 | 2,342 | 2,661 | 2,593 | 1,412 | 1,828 | 1,719 | 1,788 |
| Brazil | 425 | 549 | 623 | 717 | 939 | 1,038 | 1,381 | 1,556 |
| Turkey | 192 | 400 | 500 | 655 | 880 | 800 | 850 | 675 |
| Francophone Africa | 63 | 140 | 224 | 562 | 728 | 937 | 888 | 571 |
| Greece | 63 | 110 | 115 | 213 | 421 | 430 | 300 | 300 |
| Australia | 2 | 19 | 99 | 433 | 804 | 589 | 253 | 116 |
| *World* | **10,201** | **11,740** | **13,831** | **18,970** | **19,437** | **24,775** | **25,312** | **26,213** |
| **EXPORTS** | | | | | | | | |
| United States | 1,444 | 848 | 1,290 | 1,697 | 1,472 | 3,821 | 3,048 | 3,092 |
| Central Asia | 381 | 553 | 876 | 1,835 | 1,203 | 1,454 | 1,467 | 1,324 |
| India | 53 | 34 | 140 | 255 | 24 | 800 | 1,050 | 1,300 |
| Francophone Africa | 48 | 137 | 185 | 498 | 767 | 1,013 | 928 | 609 |
| Brazil | 152 | 220 | 21 | 167 | 68 | 429 | 300 | 500 |
| Australia | 0 | 4 | 53 | 329 | 849 | 628 | 483 | 270 |
| Greece | 33 | 0 | 13 | 86 | 244 | 355 | 243 | 223 |
| Egypt, Arab Rep. of | 346 | 304 | 162 | 18 | 79 | 100 | 100 | 125 |
| Syrian Arab Rep. | 97 | 134 | 71 | 91 | 212 | 191 | 100 | 65 |
| *World* | **3,667** | **3,875** | **4,414** | **5,081** | **5,857** | **9,801** | **8,270** | **8,247** |

*Source:* ICAC various issues.

*Note:* Bangladesh is included in Pakistan through 1970. Francophone Africa comprises Benin, Burkina Faso, Cameroon, the Central African Republic, Chad, Côte d'Ivoire, Guinea, Madagascar, Mali, Niger, Senegal, and Togo. Central Asia comprises Azerbaijan, Kazakhstan, the Kyrgyz Republic, Tajikistan, Turkmenistan, and Uzbekistan. Years refer to marketing seasons, for example, 2006 is 2006/07 (August through July).

## The Demand Side: Changing Focus of Demand and Competition with Synthetic Fibers

Between 1960 and 2005, global cotton demand grew at the same rate as the population (close to 2 percent per year), implying that per capita cotton consumption has remained steady at about 3.5 kegs per year. By contrast, consumption of chemical fibers, which compete with cotton, has increased over the last 50 years by 2.2 percent per year, causing cotton's share in total fiber consumption to decline from 60 percent in 1960 to less than 40 percent in 2005.

Cotton lint consumption is determined by the location of the textile industries. During the 1960s, Europe, the United States, and Japan were major textile manufacturers and hence major consumers of cotton lint. Gradually, however, textile industries moved to South and Southeast Asia. Today, China, India, Pakistan, and Turkey account for more than 70 percent of global cotton consumption. Key reasons for the relocation of textile industries to these countries include low wage and energy costs and the ability to deliver final goods in a timely fashion. Currently, the 10 largest cotton importers account for more than 70 percent of global cotton trade. Three major producers—China, Pakistan, and Turkey—also import cotton lint to supply their textile industries. The four East Asian textile producers—Indonesia, the Republic of Korea, Taiwan, and Thailand—accounted for more than 20 percent of world cotton imports in 2005, compared with just 3 percent in 1960. The shift of cotton consumption to Asia has been aided by the abolition of the Multi-Fiber Agreement, which, in effect, dictated the trade flows of textile products. Southeast Asia also has the highest concentration of chemical fiber production.

## Declining and Volatile World Prices for Lint

Real cotton prices have declined over the last two centuries, although with temporary spikes. The reasons for the long-term decline are similar to those characterizing most primary commodities: on the supply side, reduced production costs resulting from technological improvements, and on the demand side, stagnant per capita consumption and competition from synthetic products (for some commodities, including cotton). Between 1960–64 and 1999–2003, real cotton prices fell 55 percent, remarkably similar to the 50 percent decline in the broad agriculture price index of 28 commodities (figure 2.1). Two periods of pronounced spikes in commodity prices during that period were the oil-induced boom of the 1970s and the price collapse of the mid-1980s. The 1985 cotton price collapse was a result of a policy shift in US commodity programs (including cotton). It also reflected a policy shift in China that favored domestic cotton production.

In addition to their declining trend, cotton prices have been volatile, a common phenomenon among primary commodities. The degree of volatility, however, changed considerably during the last 40 years. Various measures of price volatility calculated by Baffes (2005) show three distinct periods: (i) 1960 to 1972, when prices were very stable; (ii) 1973 to 1984, when various measure of volatility quadrupled; and (iii) 1985 to 2002, when volatility fell by half but remained twice what it was during the period 1960–72. This conclusion is similar to findings by Valdès and Foster (2003), who looked at price variability of corn, rice, sugar, and wheat, as well as findings by Sarris (2000), who examined intra- and inter-year price variability of wheat and maize.[13]

Cotton has not been part of the recent commodity price boom (figure 2.2). Likely reasons include the fact that (i) cotton subsidies continue to depress prices (see box 2.1 on subsidies); (ii) GM

**Figure 2.1 Real agricultural price index (based on 28 commodities) and cotton prices adjusted by the MUV, 1960-2008 (1980=1.0)**

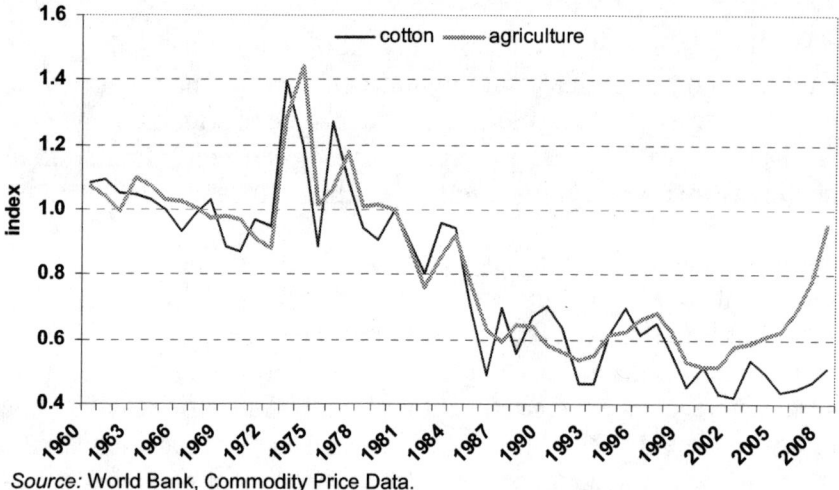

*Source:* World Bank, Commodity Price Data.
*Note:* MUV = manufacturers unit value.

cotton production is expanding, particularly in China and India, where production costs have been kept low, in part as a result of rapid expansion of GM cotton varieties; (iii) the prices of many other commodities have recently rallied because of the increasing demand for biofuels production (for example, maize for ethanol production in the United States and rapeseed for biodiesel production in the European Union); and (iv) most food commodities have responded to increased energy costs much more strongly than cotton.[14]

## External Factors Affecting World Lint Prices

Two major factors affecting African cotton trade—but for the most part not under full control of individual governments—must be underscored: (i) the impact of foreign exchange variations, that is, the relative value of the national currency with respect to the dollar; and (ii) the

**Figure 2.2 Nominal cotton prices (US $ per kilogram) and agricultural price index (Oct 2001 = 1.0), January 1985-September 2008**

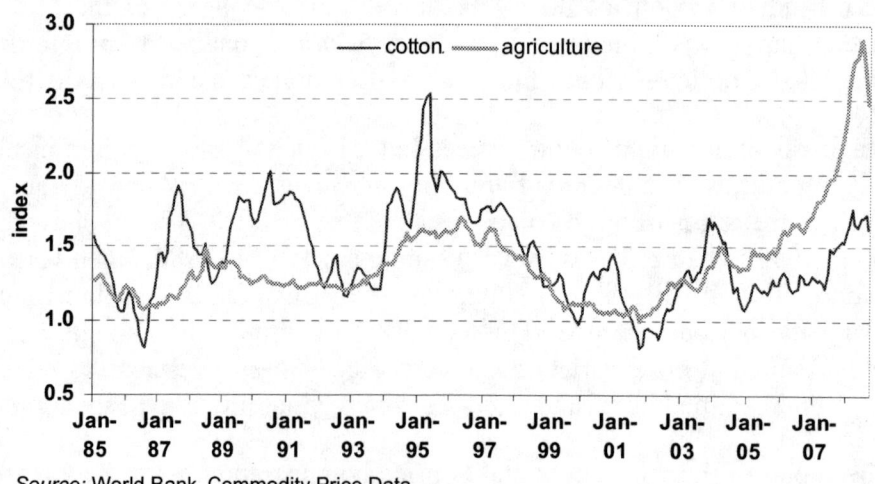

*Source:* World Bank, Commodity Price Data.

## Box 2.1 Cotton Subsidies

Cotton subsidies in the United States have a long history dating from the commodity programs of the Great Depression. The specific provisions of these programs, including the one for cotton, change with each farm bill passed by the Congress (farm bills are introduced approximately every four to five years), but their chief objective has remained largely unchanged: to transfer income from taxpayers (and to some extent consumers) to producers. The main channels of support to US cotton producers are the following: (i) *price-based payments* (also known as loan rate payments) are designed to compensate cotton growers for the difference between the market price and the target price when the latter exceeds the former; (ii) *decoupled payments* (renamed direct payments in the 2002 farm bill) are predetermined annual payments calculated on the basis of area historically used for cotton production (direct payments were introduced with the 1996 farm bill to compensate producers for "losses" following the elimination of deficiency payments); (iii) *crop insurance* is a subsidy to provide protection against weather-related crop failures; (iv) *countercyclical payments* were introduced in 1998 (as "emergency payments") to compensate producers for income "lost" because of low commodity prices. The countercyclical payments were made permanent under the 2002 farm bill. In addition to these transfers there are other publicly funded programs—among them research and extension services and subsidized irrigation. The US cotton program, which was subject to review by the U.S. General Accounting Office twice (1990 and 1995), was, and still is, very complex and expensive.

The European Union (EU) also supports its cotton producers. Between 1997 and 2007, the budgetary expenditure on the cotton sector ranged between $0.7 and $1.0 billion, implying that, on average, EU cotton producers received more than twice the world price of cotton. EU cotton producers received support even in periods of high prices—because the budgetary allocation to the cotton sector must be disbursed. For example, producers received approximately the same level of support in 1995 and 2002, although cotton prices in 1995 were twice the level of 2002. A major restructuring of the EU cotton program was undertaken under the Luxembourg Council's decision of April 22, 2004, which was based on the September 2003 proposal. Under the new program that went into effect in 2006, an estimated €700 million (almost US$1 billion) funds two support measures, with 65 percent of the support taking the form of a single decoupled payment and the remaining 35 percent taking the form of an area payment (European Commission 2003). A minor change took place in 2008 regarding the eligible base area, but this does not affect the amount of total support.

distortions created on the world market for cotton lint by the subsidies paid by some of the Organisation for Economic Co-operation and Development countries and by China to their cotton producers.

## CURRENCY EXCHANGE RATES

Because cotton is traded in US dollars (as are most commodities), cotton prices are affected by the movement of the US dollar in addition to all other factors affecting cotton prices. For example, the US dollar's strength after the East Asian financial crisis was partly responsible for the collapse of commodity prices in the late 1990s and early 2000s. Similarly, the weakness of the US dollar contributed to the commodity price boom at the middle of the 2000s.[15] However, all cotton producers are affected by the variability of the currency in which their cotton is sold. Countries examined in this report are affected by three exchange rate–related issues: (i) appreciation of the CFA franc, particularly critical for WCA countries; (ii) the Dutch disease effect in Zambia; and (iii) hyperinflation in Zimbabwe.

The CFA franc, the currency of 14 WCA countries, is linked to the euro (or to the French franc before 2000; see box 2.2 for the history of the CFA franc). Following the 1994 devaluation of the CFA franc against the French franc, WCA countries in general, and their cotton sectors in particular, gained competitiveness; production in the four WCA study countries nearly doubled over the next four years (compared with a 20 percent rise over the previous four years), driven especially by Burkina Faso and Mali. However, following the very strong appreciation in recent years of the euro against the US dollar (and consequently of the CFA franc), the competitive

## Box 2.2 The CFA Franc and Cotton in WCA

The CFA franc (CFAF) is the common currency of 14 WCA countries comprising two groups, all members of the African franc zone. One group includes Benin, Burkina Faso, Côte d'Ivoire, Guinea Bissau, Mali, Niger, Senegal, and Togo, which form the West African Economic and Monetary Union (WAEMU) and whose common central bank is the Central Bank of West African States. The other group includes Cameroon, the Central African Republic, Republic of Congo, Gabon, Equatorial Guinea, and Chad, which form the Central Africa Economic and Monetary Community (CEMAC) and whose common central bank is the Bank of Central African States (BEAC).

The CFAF was created in 1945, when France ratified the Bretton Woods agreement. At that time, the CFAF was the acronym for Franc of the French Colonies of Africa (*Franc des Colonies Françaises d'Afrique*). In 1958, it became Franc of the French Community of Africa (*Franc de la Communauté Française d'Afrique*). Today it means Franc of the African Financial Community (*Franc de la Communauté Financière d'Afrique*) for WAEMU members and Franc of Financial Cooperation in Central Africa (*Franc de la Coopération Financière en Afrique Centrale*) for CEMAC members. Initially, convertibility with the French franc (FF) was set at 0.59 CFAF/FF, becoming 0.50 CFAF/FF after the 1948 devaluation of the French franc. In 1958, two zeros were added to the existing denomination, making it 50 CFAF/FF.

During the early 1990s, it became increasingly apparent that the CFAF was overvalued. The degree of overvaluation, however, differed markedly among WCA countries. Baffes, Elbadawi, and O'Connell (1999), for example, based on a single-equation framework, estimated that during the early 1990s, the CFAF was overvalued more than 30 percent in Côte d'Ivoire while it was roughly in equilibrium in Burkina Faso. Devarajan (1999), based on a simple general equilibrium model, concluded that the CFAF overvaluation in 1993 (one year before the devaluation) was 78 percent in Cameroon, 52 percent in Togo, 39 percent in Mali, 36 percent in Côte d'Ivoire, 22 percent in Senegal, 9 percent in Burkina Faso, 3 percent in Benin, and −19 percent in Chad (that is, undervalued). For an extensive discussion of issues surrounding the CFAF overvaluation, see Hinkle and Montiel (1999). In January 1994, the CFAF was repegged to the French franc at 100 CFAF/FF and in 1999 it was linked to the euro at 656 CFAF/€, keeping its former parity with the FF.

The 1994 adjustment to the CFAF, which temporarily restored currency equilibrium in most WCA countries, coupled with the cotton price increases of the mid-1990s, induced considerable supply response in the cotton sectors of most WCA countries. For example, regional cotton production increased from 573,000 tons in 1993/94 (the year before devaluation) to 921,000 tons in just four years. For the nine years that followed, however, cotton output remained, for the most part, stagnant at 900,000 tons. Such stagnation along with the financial difficulties of the cotton companies may be a result of, in part, the likely overvaluation of the CFAF. This should not be surprising. During 2005/06, the US dollar Cotlook A Index (that is, the world price of cotton) average was roughly the same as in 2000/01. However, during the same period the CFAF appreciated from 731 CFAF/$ to 535 CFAF/$, effectively reducing the world price of cotton in CFAF terms by 37 percent.

edge that it was giving to CFA countries in the 1990s was completely reversed. Figure 2.3 depicts the CPI-adjusted bilateral exchange rate of three countries, India, China, and Burkina Faso, since January 2002. While India and China have seen little real depreciation, the CFA franc in Burkina Faso has appreciated by almost one-third against the US dollar in real terms. These fluctuations affect the performance of the cotton sectors: China's cotton production increased from 5.3 million tons in 2003 to 8 million tons in 2007. Similarly, India's cotton production increased from 3.0 to 5.3 million during the same period. Cotton production in WCA, however, declined from 0.91 to 0.79 million tons during the period.

Exchange rate movements are also an important issue in Zambia, but for a different reason (Tschirley and Kabwe 2007b). From 1996 through 2001, the Zambian kwacha slowly depreciated in real terms against the US dollar. As a result, export sectors with a significant share of costs in local currency would have been able to earn slightly higher profits, all else being equal. From mid-2002 to mid-2005, however, the kwacha appreciated more than 30 percent against the US dollar. Though this pattern may have been broadly consistent with depreciation of the US dollar more generally, the kwacha then appreciated an additional 35 percent over the next

**Figure 2.3 Real exchange rate for India, China, and Burkina Faso against the US dollar, adjusted by the consumer price indices, 2001-2007 (January 2002 = 1.0)**

*Source:* World Bank, Commodity Price Data.

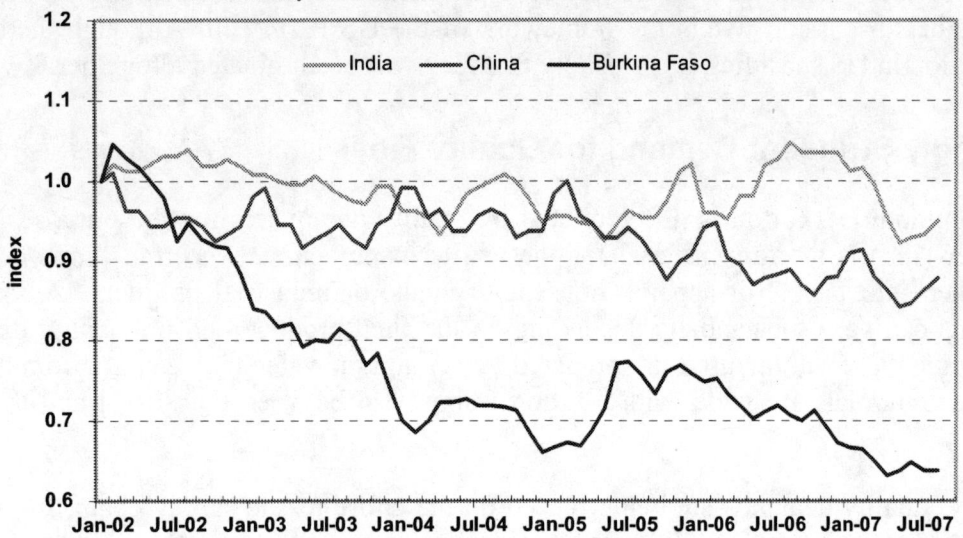

nine months, putting extraordinary pressure on its export sectors. A slight recovery in the real rate in late 2006 left it still well below typical levels from 1996 through 2002.

Zimbabwe faces exchange rate issues as well, albeit of a different kind. Since 2001, exporting companies have been required to remit a proportion of all foreign exchange receipts to the Central Bank, where they have been exchanged into Zimbabwe dollars at the official exchange rate. The exporting companies have been allowed to keep the remainder in foreign currency accounts, but since 2005 it has been difficult for companies to access and freely use the portion (currently 70 percent) kept in foreign accounts. As the ratio of the parallel exchange rate to the official exchange rate varied from 1.0 to 24.7 between January 2003 and February 2007 (average was 3.7), the proportion of their proceeds that companies were allowed to keep became an important determinant of their profitability.

**SUBSIDIES**

While there are various trade-distorting interventions in the cotton market, the most important at a global level is the domestic support given by the United States (see box 2.1). Numerous models have evaluated the impact of US cotton policies on the global market, with considerable variation in results (a summary description of these models can be found in FAO [2004]).[16] A simple average over all models shows that world cotton prices would have been between 10 and 15 percent higher without support. Applying a simple average to WCA cotton producing countries shows that these countries lost approximately $150 million annually in export earnings because of the subsidies. Cotton subsidies became a contentious issue during the current trade round under the auspices of the Doha Development Agenda. The cotton subsidy issue was further highlighted following the move by four WCA cotton producing countries (Benin, Burkina Faso, Chad, and Mali) to demand compensation for the lower prices received because of subsidies. Brazil also brought a case to the World Trade Organization (WTO) against the United States, claiming, among other issues, that subsidies reduce world prices and hence hurt Brazil's export earnings. Following the WTO's ruling, the United States removed the Step 2 payment part of its cotton program (amounting to about 15 percent of its subsidies). The WCA cotton producers, however, have not received any compensation. The fate of remaining US subsidies is currently under discussion within the framework of the US Farm Bill. Although there may be some reduction in US subsidies, it is unlikely that they will be eliminated altogether.

## Increasingly Stringent Demand for Quality Fiber

Like all commodities, cotton is differentiated by quality parameters for the purposes of trade. Cotton fiber (lint) is the raw material for the textile manufacturer who transforms cotton into yarn and then into fabric for apparel, household goods, or industrial products. Cotton quality requirements can vary substantially depending on the final product, and the quality differences affect the price that manufacturers are prepared to pay and the value they can get from the cotton lint. Price differentials are wide, with a ratio of about 1:4 between the lowest and the highest quality lint.[17]

Increasingly, quality demands are being placed on the entire textile supply chain, from the raw material to end products. Chemical fiber performance has become a benchmark for cotton spinning. The main challenge for cotton is to be able to compete with artificial fibers, mainly polyester, on both price and quality. Chemical fibers are generally easier to process, more versatile, and stronger than cotton fiber, and modern textile industry machinery requires from cotton fiber the same characteristics of cleanliness and homogeneity as those offered by artificial fibers.

In short, the increasingly stringent demand for quality cotton can be articulated in the following motto: "fiber, only fiber, but more than just fiber." The rest of this section elaborates on each aspect of this challenge.

### FIBER

Because cotton is a natural and seasonal product, its intrinsic quality (the fiber properties), its cleanliness and contamination, and the homogeneity of its characteristics can vary greatly as a result of genetic, environmental, harvesting, and ginning factors. Such variability impacts processing performance, costs, and quality throughout the cotton textile chain. Fiber properties primarily depend on varieties grown, agro-climatic conditions, and crop management practices. The cleanliness of lint refers specifically to the presence of vegetal matter other than lint, while contamination refers to the presence of nonplant matter. Both cleanliness and contamination depend on harvesting methods, storage, transport, and ginning practices.

Better fiber quality translates into better yarn quality and higher processing efficiency. Among the fiber properties, staple length has the greatest influence on spinning performance. Cotton fiber represents about 50 percent of the cost of yarn. Traditionally, the price of cotton was largely determined by factors such as staple length, grade, color, and micronaire.[18] Those factors are still the major determinants of price, but spinners today are also interested in other fiber properties that affect the quality of their yarns and the efficiency with which they can produce those yarns. As the textile industry has been striving to improve quality and efficiency through automatic high-speed machinery, new technologies place increasingly severe technical demands on textile fibers, raising the importance of other properties of cotton: strength (or tenacity), uniformity, maturity, fineness, elongation, and neps,[19] short fiber content, spinning performance, dyeing ability, and cleanliness. All else equal, spinners pay a higher price for longer, finer, and stronger cotton lint that is white, bright, and fully mature.

The most commonly produced and traded cotton lint variety in the world belongs to the species *Gossypium hirsutum*, which is also known as upland cotton. Extra long staple cotton used for producing very fine yarns come from another species[20] and accounts for less than 5 percent of world cotton trade.

Following the global trend toward improving yarn quality, the market share of medium and higher grades is rising, while the share of shorter ("coarse count") upland cotton is declining. Medium and higher grades of upland cotton now account for an estimated 75 percent of world trade, or some 7 million metric tons, and are typically used in ring spinning.

The recognized benchmark for international cotton prices, the Cotlook A Index, is based on the representative offering price for a "basket" of the medium grade cotton most commonly traded internationally. Those price quotations refer to a common quality, contractual, and geographical[21] basis (table 2.2). Lint of this quality is typically used in ring spinning for the production of ring spun carded yarns.[22] The fastest growing and most remunerative market for upland cotton is for higher grades and finer cotton used for producing ring spun combed yarns[23] for the woven and knitted apparel sector. In that segment, the modern high-speed machinery requires better fiber characteristics to operate at maximum efficiency and spin high quality yarns.

As shown in table 2.2, the fiber properties of most African upland cotton lie between these two levels, superior to Cotlook A Index specifications but not always reaching those needed for ring spun combed yarns.

In addition to requiring longer, cleaner, whiter, brighter, stronger, and finer fiber, this higher segment of the market ( for ring spun combed yarns) is more demanding of other fiber properties, such as elongation and neps. It also demands a lower variance in fiber properties, notably greater uniformity of length and lower short fiber content.

**Table 2.2 Fiber Properties of the Cotlook A Index, Typical African Upland Cotton, and Top Quality Lint for Combed Yarns**

| Fiber property | Cotlook A Index | Typical African upland cotton | Lint for ring spun combed yarns |
|---|---|---|---|
| Grade | Middling – white | Strict low middling to good middling | Strict middling — white |
| Staple length | 1 3/32 inches (27.8 mm) | 1 1/6 to 1 3/16 inches (27 to 30.2 mm) | ≥ 1 1/8 inches (28.6 mm) |
| Micronaire | 3.5 to 4.9 | 3.5 to 4.5 | 3.8 to 4.2 |
| Fiber strength | 25 to 30 grams per tex | 27 to 32 grams per tex | more than 30 grams per tex |

*Source:* Authors' estimates based on interviews.
*Note:* A tex unit is equal to the weight in grams of 1,000 meters of fiber. The strength reported is the force in grams required to break a bundle of fibers one tex unit in size.

## Only Fiber

Cotton prices are not solely determined by intrinsic fiber properties and lint cleanliness. Contamination of lint by nonvegetal foreign matter is the most serious problem confronting cotton spinners around the world. Contaminated cotton causes disruptions in the spinning process, which increases the cost of spinning and reduces the quality of yarn and end products. There are no cost-effective means of removing contamination once it is present in yarn or fabric. As a result, contamination leads to the downgrading of end products or even to rejection of an entire lot.

Cotton that is contaminated, or that is suspected of being contaminated because of its origin, can only be sold at a substantial discount to compensate the user for inspecting and cleaning the cotton before spinning. Price differentials for cotton with the same fiber characteristics range from 5 percent to 30 percent, depending on the degree of perceived contamination by extraneous matter, stickiness, and seed coat fragments. These discounts are usually applied indiscriminately to all cotton originating from an area or a country considered to be affected by contamination.

Contamination by foreign matter is more serious with handpicked cotton. Seed cotton picked by hand is cleaner, and the fiber obtained has fewer neps and a lower short fiber content than cotton picked by machine, which must be cleaned more vigorously because it has more vegetative residues. Handpicked cotton should therefore normally be purchased at a premium over machine picked cotton. However, handpicked seed cotton often gets contaminated during picking, storage, handling, or transport, and the presence of foreign matter in the fiber offsets the theoretical advantage conferred by manual picking. Because contamination of raw cotton by foreign matter is the main concern for quality yarn and fabric producers, spinners tend to prefer machine picked cotton to handpicked cotton. As a result, handpicked cotton has lost its advantage and now trades at a discount to machine picked cotton. The elimination of contamination thus stands out as the first priority for quality improvement in SSA.

## More than Fiber

Along with fiber characteristics, other criteria, such as reputation and other marketing factors generally not included in contracts, can have a lasting influence on cotton prices. However, nonquality premiums and discounts are hard to quantify because each shipper and spinner may have different opinions on a specific growth or origin.

Pricing of lint is significantly influenced by the way cotton is marketed and shipped. The spinning industry today is especially concerned about consistency in shipments. Customers require homogeneous and reliable year-round shipments, with consistent cotton characteristics, standardized bales wrapped in cotton cloth, and bale per bale instrument classification data. Because some countries can offer bale per bale Standardized Instrument for Testing of Cotton data, the lack of reliable cotton quality data on each bale negatively impacts the price of cotton that is classified manually. The homogeneity of deliveries depends on seed cotton grading, lint classification, and bale allotments.

In marketing, perception may take a long time to catch up with facts. Trust and reputation matter in the cotton business and the market rewards origins and shippers that have strong records of delivering according to quality standards and with consistency, while respecting contract terms.

Premiums and discounts attached to internationally traded cotton derive partly from the reputation of national origins.

## Lint Marketing Strategies: The Role of the International Cotton Merchant

Until the mid-1980s, most lint produced in Africa was sold by national cotton companies and marketing boards to international merchants or to spinners through commissioned agents. Today, two types of companies supply lint to the world market out of SSA: independent ginners sell lint to international cotton merchants, while ginning companies affiliated with such merchants ("affiliated ginners") sell lint to or through their mother companies.[24] International cotton merchants thus play a leading role in the marketing of African lint. They buy cotton from independent ginners or receive it from their affiliated ginners, sell it to textile mills or other merchants, hedge price risks, and arrange shipments.

Affiliated ginners are present in all countries in this review except Cameroon and Mali, which continue to operate national monopoly sectors. Most lint from SSA is handled by independent ginners. These include very large companies—the national monopolies of Mali and Cameroon and the former national monopolies in Burkina Faso and Benin (SOFITEX and SONAPRA, respectively)—and smaller private companies, most in ESA. As a general rule, these independent ginners have little knowledge of the world cotton market, very limited ability to use risk management tools, and receive very little feedback from the end users. International cotton merchants are thus in a strong negotiating position when dealing with independent ginners.

In WCA, forward sales contracts at fixed price have been used extensively for decades, primarily as a way to secure input and crop financing. Sales are contracted in euros per kilogram free on board, offsetting the exchange risk. Ginners usually base their prices on the Cotlook A Index,[25] valued at the forward exchange rate for the shipping period considered. Forward sales at fixed price in euro per kilogram are an effective way of mitigating risks, although high percentages of forward sales increase the risk of not being able to deliver the contracted quality and may lead to oversold situations. Smaller independent ginners in ESA are generally not in a position to guarantee the volume and the quality of their production before it is ginned, are not able to store lint for an extended period, and therefore seldom engage in forward sales. They primarily deal with price and exchange rate risks by adjusting their buying price over the course of the season and by selling the lint as it is ginned.

Merchants carefully select sellers to guarantee contract performance. Large parastatal companies in WCA are considered reliable by merchants, defaulting only in good faith, while private independent ginners have mixed reputations. Merchants generally consider it much easier to purchase cotton in WCA countries than from independent ginners in ESA countries because offers are not spread over numerous small trading companies,[26] and the volume is sufficient to ensure year-round shipments, while quality standards are relatively consistent. In contrast, small independent ginners in ESA are generally not in a position to guarantee consistent or year-round shipments and large volumes.

In addition to fixed price contracts, some basis pricing is done "on call" the Cotlook A Index, "on call" the African franc zone quotation in *Cotton Outlook*, or "on call" New York futures. On call pricing means that the buyer (or seller) agrees to a volume and delivery date, but that the price will be fixed at a later time. Major fluctuations in the basis (arithmetic difference) between spot prices and the New York futures can present challenges to hedging of African cottons.

Competition among buyers and among ginners of seed cotton within a producing country may increase the producer price, especially when there is overcapacity in ginning. Among the countries included in this study, this practice is most common in Tanzania and Uganda. Yet the immediate impact of increased competition between ginners and exporters is to put pressure on the selling price of lint because buyers (merchants and spinners) always take advantage of competing offers from several sellers to buy from the cheapest. A national monopoly (state-owned or private) is generally in a better position to protect its selling price.

Affiliated ginners are generally more aware of market demand and have less exposure to international price fluctuations than independent ginners because a part of the market risk is taken by the mother company, which has the ability to hedge price and exchange rate risks.

## Valorization of By-Products: Markets for Cottonseed Oil and Cake

The ginning process separates the cotton lint from the seeds. In all of the study countries, ginning companies buy the raw cotton and own both the lint and the seed. In all countries but Zimbabwe, ginners treat a small proportion (typically less than 10 percent) of the seeds and pack them for distribution back to farmers as planting seed. Ginners sell the remainder[27] for processing into oil and cake or, in a few cases, the seeds are processed directly by the ginning company.[28]

The value of lint obtained from a ton of seed cotton is three to four times the combined value of the oil and cake that are derived from processing the seeds. For this reason, oil and cake markets are often neglected in the analysis of African cotton sectors. Yet in some of the study countries, the revenue from seed sales more than covers the cost of ginning. In Cameroon, SODECOTON has so far avoided the losses that have affected the sector in other WCA countries, in part because of cross-subsidization from its profitable, integrated oil business. In Tanzania, more than a third of cotton ginners now have an integrated oil processing business. Some ginners see oil processing as their core (and most profitable) business, with seed cotton purchase and ginning feeding into this.

The importance of the markets for cotton seeds, oil, and cake is likely to grow in the near future, for two reasons. Internally, as margins in lint production get tighter, cotton companies are likely to look for more ways to improve their profitability, by paying greater attention to the value they can obtain from these coproducts. Externally, increased demand for biofuels, and potential rapid growth in southern European dairy sectors (which will increase demand for protein concentrates), may open new markets and strengthen prices for cotton ginning coproducts. Depending on policy and investment climates in cotton producing countries, these changes could drive more local competition for coproducts and alter the relationship between farmers and ginning companies over time.

### INTERNATIONAL TRADE IN COTTONSEED OIL

It is estimated that about 5 million tons of cottonseed oil is produced worldwide per year. This amount is similar to the production of groundnut, coconut, and palm kernel oil, but well behind palm oil (38 million tons), soybean oil (37 million tons), rapeseed oil (18 million tons), and sunflower oil (11 million tons). Moreover, while a large proportion of the four leading oils is traded internationally, less than 10 percent of global cottonseed oil production is traded.

World cottonseed oil prices averaged $1,600/ton during 2007/08, up from $800/ton during 2006/07 (October to September) These price increases are similar to those of other edible oils. For example, the World Bank's fats and oils price index increased 51 percent from 2006 to 2007 and 44 percent from 2007 to 2008. Cottonseed oil used to be traded with a large premium over the four leading oils. However, the premium has almost disappeared lately because of the high biodiesel demand for rapeseed oil, which has also put pressure on soybean and palm oil prices.[29]

## DOMESTIC OIL AND CAKE MARKETS

Virtually all cottonseed oil produced in the countries studied here is consumed domestically. Because all countries are net edible oil importers (in most cases palm oil), cottonseed oil is an import competing crop. Table 2.3 estimates the proportion of edible oil consumption that can be supplied from cotton seeds in each of the nine study countries. This ranges from 50 percent or more in Benin, Burkina Faso, and Mali to less than 10 percent in Mozambique, Tanzania, and Uganda.

Cottonseed cake is an oil processing by-product sold as livestock feed. It is rarely traded internationally because of its low value-to-weight ratio. Demand for cake, therefore, depends heavily on the size and degree of commercialization of the local livestock (cattle, poultry) industry.

**Table 2.3 Summary of Indicators of Valorization of By-Products**

| Country | Average national seed cotton production, 2001–06 (tons) | Cottonseed oil production as % of national oil consumption[a] |
|---|---|---|
| Benin | 339,500 | 53 |
| Burkina Faso | 557,833 | 57 |
| Cameroon | 242,966 | 18 |
| Mali | 488,281 | 50 |
| Mozambique | 72,178 | 6 (potential) |
| Tanzania | 235,000 | 8 |
| Uganda | 78,410 | 4 |
| Zambia | 160,000 | 20 |
| Zimbabwe | 246,350 | 27 |

*Source:* Calculations based on information given in the country studies.

a.  Estimated from the quantity of seed available for crushing (2001–06 average), after subtraction of seed retained for redistribution to farmers, using an oil outturn of 18 percent and an average annual oil consumption of 7 kg/person.

# SECTION II
# HISTORICAL BACKGROUND AND
# CONCEPTUAL APPROACH

# Chapter 3: Historical Background and Recent Institutional Evolution of African Cotton Sectors

*Nicolas Gergely and Colin Poulton*

This chapter provides the historical background and recent evolution in the cotton sectors of the study sample. The presentation is structured according to the two subsets of countries—West and Central Africa (WCA) and East and Central Africa (ESA)—because the countries in each group feature a relatively common history, at least in prereform models.

## West and Central Africa

Cotton was introduced in most Francophone countries of West and Central Africa (WCA) in the last decades of the colonial period, as part of a broad policy aiming to supply the French textile industry with raw material. To that end, the French government created a dedicated parastatal company, CFDT (Compagnie Française pour le Développement des Fibres Textiles). CFDT was entrusted with developing cotton cultivation as an integrated supply chain—from the provision of input to farmers to the marketing of lint—in countries of Francophone Africa. After independence in 1960, CFDT continued to operate through various country or subregional branches until the mid-1970s, when these branches were turned into national companies with a majority of shares belonging to governments, and a minority retained by CFDT.[30] Most of these companies[31] entered into long-term technical assistance contracts with CFDT. These companies were usually granted a legal monopoly on the purchase and processing of seed cotton and on lint marketing, and were obliged to purchase all seed cotton production at a fixed price set by the government. Ginning was based on large units using saw gin equipment.

During the 1980s, the national cotton companies expanded their activities considerably, often with the assistance of internationally funded development projects: they increased ginning capacity, further developed input credit schemes, invested in transport for seed and lint cotton, and created their own extension services to disseminate technical packages.[32] As in the past, the national cotton companies continued to guarantee purchase of the crop at a fixed, panterritorial price announced before planting. In some cases (Mali, and, to some extent, Cameroon), the companies were also given responsibility for rural development activities in the cotton areas. Cotton production grew rapidly as a result of these investments, based on an increasing number of cotton farmers and increased farm yields and ginning outturn ratios. Lint quality also improved. Yields increased dramatically in most countries until the mid-1980s, thanks to intensified use of fertilizer (made possible through input credits), development of animal traction, and development of new varieties with higher yield potential, as well as higher ginning outturn ratios. Most varieties were developed in cooperation with IRCT (the French public Cotton and Textile Research Institute, which later merged into CIRAD, the French Agricultural Research Centre for International Development). Meanwhile, to cope with increasing seed production, large-scale cotton seed processing units designed to supply domestic markets with quality refined oil were built, often as part of the cotton companies (in Cameroon and Mali).

## STRENGTHS AND WEAKNESSES OF THE WCA MODEL

During the three decades following independence (1960–90), cotton development in WCA was widely regarded as a success story, with impressive and steady growth and outreach to nearly every farmer in cotton zones. However, the rapid growth of the cotton companies put increasing strain on their management capacities, and most of them lacked adequate control and supervision systems. Aware of the risks, most WCA governments introduced in the 1980s a new monitoring instrument called performance contracts, to be negotiated between the state, the cotton company, and, in some cases like Mali in the early 1990s, with the cotton farmer organization. The overall objective of the performance contracts was to make cotton companies more accountable to governments and stakeholders. Specific objectives included untangling commercial activities from public service activities and securing separate funding for these activities, as well as establishing financial performance targets for the cotton companies, based on standard costs (*barêmes*).[33] These performance contracts were implemented in Burkina Faso, Cameroon, and Mali, from the mid-1980s to the end of the 1990s. In practice, the contracts proved difficult to monitor in the absence of a strong and independent reporting system; cotton companies were reluctant to provide information, would often argue that changes in the economic environment justified their performance, and would still turn to the government to cover losses. In the end, the era of performance contracts failed to deliver any significant and long-lasting improvements in the governance of the cotton sectors.

The need for deeper structural reforms began to appear for the first time at the end of the 1980s, when WCA cotton sectors faced financial difficulties as a result of the cotton companies' poor cost efficiency, declining world prices, and an overvalued local currency. After the 1994 devaluation of the CFA franc, which boosted both production and cotton company profits, the need for reform was perceived as less urgent. World prices also surged from 1994 to 1996. The immediate postdevaluation period thus saw rapid production growth and high profits for the cotton companies, but often lax management practices, resulting in high cost structures. Farm yields had also begun to stagnate by this 1996. When world prices declined at the end of the 1990s, cotton companies again faced serious financial difficulties, which were aggravated when the CFA franc began to appreciate against the US dollar in 2001.

It can be argued that the WCA cotton sector organizational model, long regarded as successful, became a victim of its own success in the post devaluation period (since 1994) as a result of the following constraints:

- The continuous development of cotton made WCA economies, particularly in Sahelian countries, heavily and increasingly dependent on the cotton sector. As a result, cotton farmers and other stakeholders in the cotton sector exercised considerable political and socioeconomic influence in rural areas, and management of the cotton companies became subject to increasing political interference. At the same time, the companies themselves gained economic weight and political influence, which made them even more difficult to control.

- Decisions by governments and by cotton companies became indistinguishable, and were often driven by short-term political considerations rather than by the need to ensure long-term sector sustainability.

- Because of the considerable income accumulated during the postdevaluation period, politicians were increasingly tempted to exert pressure to extract resources from cotton companies, either to finance public expenditures or for private gain.

- Finally, cotton companies failed to introduce the sophisticated management tools required for such large-scale and complex enterprises (SOFITEX in Burkina Faso and CMDT in Mali are the largest cotton companies in Africa), thus leading to growing inefficiencies and lax management control.

## CHANGES MADE SINCE THE END OF THE 1990S

By the end of the 1990s, the repetition of financial crisis among cotton companies created a strong feeling in many stakeholders that the sectors needed reform. However, in most of the region there was also a clear consensus that the single-channel relationship between producers and the cotton companies was necessary to ensure a sustainable input credit system and to guarantee intensive cropping practices and should therefore be maintained, at least at a regional level.[34] This position considerably reduced the options for liberalization and reform. Therefore, unlike in ESA countries, little structural change was effected in the cotton sectors of the reviewed WCA countries, except in Benin. At the same time, incremental change was brought to the existing single-channel model. These changes pertained mainly to (i) the development of farmer associations and their progressive involvement in the delivery of critical services and functions, (ii) the entry of private actors in ginning or input supply activities (Benin, Burkina Faso), (iii) the tentative and often partial withdrawal (in a limited number of countries) of the government from the management of the cotton sector and the parallel empowerment of cotton sector "inter-professional committees" (IPCs), and (iv) the introduction of producer price-setting mechanisms that attempted to ensure a better link to world prices. The extent of change to the original system varies substantially from one country to another in the selected sample, and in WCA in general.

***Empowerment of farmer associations.*** The first attempts to build farmer organizations began in the mid-1970s in Burkina Faso and Mali, followed by Cameroon and Benin. These associations were originally viewed by cotton companies as a means to cut their costs by transferring some functions to farmers (in particular, primary collection of seed cotton and distribution of input and seeds), and as a way to secure repayment of input credit through mutual guarantee.

The first generation of associations lacked internal cohesion and their performance was generally disappointing. They were replaced in the 1990s by smaller associations that were legally recognized and exclusively involved in cotton. In parallel, regional and national unions of associations were built up with the support of the donor community: FUPRO (Fédération des Unions de Producteurs du Bénin) was created in Benin in 1993, UNPCB (Union Nationale des Producteurs de Coton du Burkina) in Burkina Faso in 1998, and OPCC (Organisation des Producteurs de Coton du Cameroun) in Cameroon in 2000, while the process is still under way in Mali. To increase their involvement in the management of the sector, the national farmer association has been given a 20 percent share in the capital of the three privatized cotton companies in Burkina Faso, and the same move is being considered in Mali and Cameroon within the privatization process. In Mali, Benin, and Cameroon, responsibility for input supply (through competitive bids) is being transferred to farmer associations and their unions. In Benin, Cameroon, and Burkina Faso, responsibility for extension services, particularly in the field of

25

farm management, is currently being taken over by farmer unions. Ultimately, it is hoped that these more focused associations will develop the technical capacity and cohesiveness to become equal partners with the cotton companies in a balanced comanagement of the cotton sector. If happens, the government could more effectively reduce its direct involvement.

***Entry of private actors in ginning activities.*** Privatization of the cotton companies has been, in all countries, strongly advocated by a number of development partners, with the objectives of (i) providing cotton companies with clear managerial leadership, (ii) improving management practices and cost efficiency, (iii) reducing the risks of political interference, and (iv) creating smaller and more manageable enterprises. However, the involvement of the private sector has so far remained limited. The privatization process has been long and difficult. It has had to overcome the reluctance of the established cotton companies, and to be accompanied by the design, in the absence of a clear reference model, of mechanisms to ensure that delivery of critical services and functions to farmers would be preserved. In the sample of reviewed countries, only Benin (in 1995) and Burkina Faso (in 2004) have so far permitted the entry of private investors, without, however, allowing them to compete for the supply of seed cotton. Each cotton company has its exclusive zone in Burkina Faso; seed cotton is allocated administratively to cotton companies, at a fixed price, in Benin.[35] In Mali, privatization of CMDT was originally scheduled to take place in 2004, but was postponed to 2008. In Burkina Faso, the scope of the privatization process was limited by the fact that the two private cotton companies represented less than 15 percent of the country's total cotton production and SOFITEX (in which the government retains a 35 percent share) remains by far the largest ginner. In Benin, the main ginner, SONAPRA, is still a parastatal company and accounts for about 50 percent of seed cotton ginned in the country.

The expected benefits of privatization have not yet materialized, in large part because of the complicated financial situation of the sectors. In Burkina Faso, privatization had the potential to bring new investments and new partnerships with international traders but the financial crisis the cotton sector has experienced since 2005 is threatening progress. The impact of private sector entry on cost efficiency has been limited so far, probably as a result of the absence of real competition. Privatization efforts have also not clearly reduced political interference, as illustrated by the failure to reduce producer prices in response to falling world prices. In Benin, the outcome of reform clearly fell short of expectations, and resulted in a sharp decline of the sector's performance for a number of reasons: new ginners were local and often inexperienced businessmen attracted by short-term returns but without long-term development strategies, the coordination mechanisms were not really enforced, and the government played an ambiguous role with respect to vested interests.

The next country to privatize will be Mali, where the government announced in May 2008 that CMDT would be sold to private companies, each with a regional monopoly. In Cameroon, the privatization of the cotton company SODECOTON is on the agenda, but no timetable has been agreed on because of the mixed attitude of some stakeholders, who fear that it might endanger the positive role that cotton has played in rural development and social stability in the northern region of the country. Other privatizations are also planned for the short term, in Benin (SONAPRA) and Chad (COTONTCHAD). These new privatizations should build on the experience of Burkina Faso. However, the likelihood of attracting strong and professional investors, given prevailing market conditions and financial difficulties, is a concern.

***Evolution of government's role and empowerment of IPCs.*** The creation of IPCs to take over monitoring and coordination responsibility in the cotton supply chain has been viewed as a way to remedy the deficiencies of the traditional state-controlled model, and as complementing the strengthening of farmer unions. Progress in operationalizing IPCs has been mixed:

- Benin created the first IPC in 1999. The body was given a legal mandate but has been struggling to exercise its power since it was created, because final decision-making remains with the government.

- Burkina Faso created its IPC in 2006 and empowered it to regulate relationships between stakeholders in the sector, especially for the funding and provision of critical functions (extension, research, and road maintenance) and decisions on producer prices. These responsibilities were previously exercised by the government. The capacity of the Burkina Faso IPC to effectively manage the sector remains weak, as revealed by the recent financial crisis, by the marginal role of the two private cotton companies in decision-making processes, by the residual influential power of the government (in particular, through SOFITEX and local parastatal banks), and by the absence of regional coordination bodies in each of the concession zones.

- An IPC is scheduled to be created in Mali in 2008 (after the privatization of CMDT).

- In Cameroon there is only one cotton company and one farmer organization. The creation of an IPC has not, up to now, been deemed necessary to achieve comanagement of the supply chain, but the two stakeholders have reached a high degree of cooperation in the decision-making process on all sector issues.

***Evolution of producer pricing mechanisms.*** One major drawback of the monopoly system for ginning companies is that it has been accompanied by the setting of fixed producer prices. When producer prices are fixed before planting, ginning companies take on great risk. The extent of this risk was fully revealed in the 2003/04 season, during which world prices dropped nearly 30 percent, creating trading losses for the ginning companies. All WCA countries have had to face this issue in recent years.

With the building up of producer organizations, all WCA countries in the study shifted from prices administered by the government to prices set jointly by the cotton companies and producer organizations. In all of these countries, the price mechanism was, at least until the 2004 crisis, linked to a stabilization fund designed to support producer prices when the world market was low, and to be replenished when the world market was high (by paying farmers lower prices than could otherwise be paid). The rationale for these funds was to avoid dramatic drops in producer prices and to limit market risks for cotton companies. These support funds functioned well in Burkina Faso and Cameroon until 2004. Since then, however, they have fallen victim to the unsustainably high prices agreed to among cotton companies and producer unions. In Cameroon, the fund was exhausted by 2006, though it was sufficient to cover sector losses and so did not draw on the government budget. In Burkina Faso, the fund has been depleted and could not cover the deficits during the 2005 and 2006 seasons. In Mali and in Benin, the funds were exhausted before 2004, and therefore could not be used when world market prices started to fall. The losses in the cotton sectors of those countries were covered by direct government subsidies.

To remedy this situation, the Burkina Faso IPC adopted in 2006 a new price-setting mechanism, based on a formula linked to world market trends. This system represents a move in the right direction, but was not correctly applied for the 2006/07 season, resulting in additional losses for the cotton companies. In Mali, a new system was adopted for the period 2005 to 2008 based on a conservative initial producer price[36] and on sharing of the actual selling price, at the end of the season, between producers and the cotton company. This system reduced the losses of the cotton company, yet was unable to completely eliminate the losses during the 2007 season.

Establishing price mechanisms that are acceptable to farmers and sustainable for cotton companies (while providing incentives for them to be more cost efficient) appears, therefore, to be one of the major challenges for WCA cotton sectors in the near future. This is especially important if other features of the model (announcement of the producer price before the planting season, panterritorial pricing, obligation to purchase all seed cotton offered) are not reconsidered, that is, as long as no additional steps are made in liberalizing the sector.

WCA cotton sectors are currently engaged in discussions, with the assistance of donors (in particular, Agence Française de Développement), on a possible link between price-setting mechanisms inspired by the new Burkinabe system (that is, based on world market trends), and a national "smoothing fund" backed by a regional refinancing facility (at the level of the Economic Community of West African States). This fund would guarantee the payment of the producer price, but, unlike the former stabilization funds, would be operated and managed by a bank (under monitoring by the IPC) according to predefined rules. The expectation is that this approach will avoid manipulation of the rules or misuse of the fund.

## CONCLUSION

Though limited structural change has taken place in WCA, incremental institutional and organizational changes have been significant. The strengthening of farmer organizations has paved the way for active involvement in critical activities, such as input supply and extension, in which they have a potential comparative advantage, and has also created the possibility of active comanagement of the cotton sectors by farmers and ginners through the IPCs. The reform of pricing systems has been difficult but it was necessary to make producer prices more connected to world market prices. The introduction of private cotton companies in Burkina Faso has shown the ability of national single-channel systems to shift toward local concession systems without disrupting input credit supply, though the potential advantages of such a change have not yet materialized because of other constraints.

## East and Southern Africa

In East and Southern Africa (ESA), there is perhaps greater heterogeneity in the historical experience of cotton sector development compared with WCA, but there are important common threads across countries over time. One important contrast with WCA is the role assumed by the cotton sector in broader rural development. In WCA, colonial governments and then independent states made cotton an engine of development and organized the *filière* (supply chain) to serve that objective. By contrast, cotton cultivation in ESA typically had its origins in commercial or missionary activity, with the government assuming a greater role over time. Mozambique is the one exception to government control in ESA. In Mozambique, the Portuguese colonial regime treated cotton as a strategic commodity, but the independent government was unable to maintain that commitment in the face of war and economic collapse.

The cotton sectors in Uganda and Tanzania have always been based on smallholder production, spurred by the colonial requirement that smallholder households pay taxes. In the early twentieth century, Asian businessmen dominated seed cotton purchase and ginning, while governments assumed responsibility for research and extension, seed multiplication, quality control, and lint export. Uganda was Africa's largest lint exporter until the beginning of the 1970s. Until the mid-1950s, the Uganda Lint Marketing Board was also responsible for the export of Tanganyikan lint.

The cotton sector in Mozambique was also based on smallholder production for most of its history, with the exception of 1965–75, when growing opposition to the colonial government in the north (the cotton growing heartland) prompted the government to promote production on large European-controlled farms. By contrast, large-scale farmers of European origin drove early sector development in Zimbabwe. At least until 1980, the large-scale farmers had the political power to advocate for the establishment and maintenance of research and marketing systems to support their production activities.

## ESA COTTON SECTORS IN THE POST-INDEPENDENCE YEARS

From 1960 to 1990, two main changes occurred in ESA cotton sectors. First, countries achieving independence transferred control over the sector, with the government (or government-controlled organizations) playing an increasing role in seed cotton purchase and ginning at the expense of the private sector. The purported reason for these changes was typically to support smallholder cotton farmers. At the same time, performance declined seriously in all countries except Zimbabwe.

In both Uganda and Tanzania, regional cooperative unions replaced Asian businessmen as buyers and ginners of seed cotton. The cooperative movement started as a member-driven phenomenon, but politicians soon exerted government control. In Uganda, cooperatives were given monopoly rights over seed cotton purchase and ginning in 1969, with the Lint Marketing Board handling the marketing of lint and seed and regulating the cooperatives. In Tanzania, cooperatives displaced Asian businessmen during the 1960s, initially with farmer support, then through force of law. At the same time, the government attempted to control the powerful Victoria Federation of Cooperative Unions (seen as an alternative center of power to the ruling party) by replacing it with the Nyanza Cooperative Union. State-imposed cooperatives performed poorly in a number of the country's main cash crops, thus cooperative unions were abolished in 1976 and a parastatal Tanzania Cotton Authority assumed responsibility for crop purchases from village-based cooperative societies.

Cotton production experienced a precipitate collapse in both Uganda and Mozambique in the mid-1970s. With the seizure of power by Idi Amin, lint production in Uganda plummeted from 78,000 tons in 1972 to just 14,000 tons in 1976, undermined by poor policy and escalating costs and mismanagement at the cooperatives and the Lint Marketing Board. Similarly, with independence in Mozambique in 1975, seed cotton production fell from a peak of over 140,000 tons in 1973 to below 40,000 tons in 1976. Production by commercial farmers collapsed to around 20 percent of its immediate pre-independence peak, and smallholder production declined sharply, discouraged by disastrous central planning policies. With the outbreak of civil war in both countries, production fell further, to lows of 5,200 tons of seed cotton in Mozambique in 1985 and 2,000 tons of lint in Uganda in 1987.

In both Tanzania and Zambia, government mismanagement of the cotton sector led to mounting debt and eventually to delayed payments to farmers. However, the impact on production was nowhere near as disastrous as in Uganda or Mozambique. In Zambia, sector development was the responsibility of the parastatal Lintco from 1977 onward. Annual production rose from around 3,000 tons during 1974–76 to a peak of over 60,000 tons in 1988, then trended down to 30,000 tons by 1994. Lintco debts also increased to the point where the government decided to privatize it.

Cooperative unions were reinstated in Tanzania in 1984 as part of an economy-wide reform. Cotton production, which had declined steadily under Tanzania Cotton Authority management, began to recover, and reached record levels in 1991 and 1992. Production during the latter year was over 300,000 tons of seed cotton, a level that would not be reached again until 2004. The cooperative system delivered some credit to farmers and, until at least the late 1980s, Tanzania maintained a reputation for good quality lint. However, the inefficient restored cooperative unions required increasing financial assistance from the central government (mostly as guaranteed loans from government banks, despite nonrepayment of previous loans because of trading losses). As mismanagement and shortages of funds caused cooperative unions to take quality less seriously, Tanzania's reputation for lint quality began to decline (before the impacts of liberalization).

The good performance of Zimbabwe's cotton sector during this period stands in contrast to that of the other ESA countries in the study. Production expansion during the 1960s was founded on two research breakthroughs: the introduction of the high-yielding Albar 637 seed variety in 1959–60 and effective chemical control of red bollworm. Production levels were maintained during the 1970s despite the escalating liberation war. Half of the governing board of the Agricultural Marketing Authority (AMA)—set up in 1967 to coordinate the activities of the Cotton Marketing Board (CMB) and other major parastatals—was made up of representatives from the Rhodesian National Farmers' Union. In 1976, the AMA began to announce generous guaranteed minimum cotton prices before planting.

Following independence in 1980, activities of the CMB were reoriented toward meeting the needs of new, smallholder cotton producers in so-called communal areas. The number of buying posts in such areas was greatly increased and efforts were made to provide smallholder farmers with extension advice, while new seed varieties suited to production conditions in communal areas were developed. In addition, expansion of smallholder cotton production was supported by loans from the parastatal Agricultural Finance Corporation. Nevertheless, commercial farmers still accounted for 60 percent of national production in 1988.

Commercial farmers in Zimbabwe began to exit cotton for more profitable alternatives in the late 1980s and early 1990s. The CMB responded in 1992 with the introduction of a credit scheme designed to assist smallholder farmers in expanding their cotton production. By the time of sector liberalization in 1994, smallholders accounted for 60 percent of production; their share had risen to almost 90 percent by the onset of the fast-track land redistribution program in 2001.

## COTTON SECTOR REFORM AND EVOLUTION IN ESA

Mozambique was the first of the countries in this study to embark on thorough reform of the cotton sector. In 1986, the first of four joint venture companies, a collaboration between the government of Mozambique and Lonrho, was established and given exclusive rights to run a cotton concession area in Cabo Delgado province. Because the country was still fighting a civil war, developing cotton production entailed high costs (including some infrastructure investment and hiring private militia to protect company assets), so local monopoly rights over cotton purchase were considered necessary to give some assurance of a return to investment. However, the sector continued to be based on such local monopolies, and some have argued that the open-ended nature of the concession rights is at the root of subsequent disappointing performance.

The first concessions were granted to fully private companies in Mozambique in the 1990s; by 2002 there were at least 12 companies promoting cotton, all within the concession system. However, there were two periods—in the mid-1990s and around 2000—when new entrants began buying in concession areas, effectively challenging the concession system. On the one hand, this reflected dissatisfaction with the performance of some of the existing concession companies. On the other, it reinforced any reluctance that these companies had to invest in better service delivery, because they could not be sure of capturing the returns. Both issues were eventually resolved by granting new concessions to the more powerful new entrants. In addition, the entrance of several international companies has raised hopes that sector performance will begin to improve. Notions of liberalizing the market receded, replaced in 2007 with proposals (not yet implemented as of mid-2008) to more rigorously monitor the performance of concession companies and to re-award concessions, perhaps on a five-year cycle.

The other four ESA countries in the study (Tanzania, Uganda, Zambia, and Zimbabwe) all liberalized their cotton sectors during 1994/95, when world prices were near an all-time high.

The initial structure of the liberalized sectors mirrored their preliberalization organization. In Tanzania and Uganda, where ginning was historically dominated by roller gins (cheaper and with few economies of scale) in the hands of decentralized cooperatives, a large number of private buyers and ginners entered the sector. Both countries quickly grew to more than 30 seed cotton buyers. In Zambia and Zimbabwe, where ginning was historically dominated by saw gins (larger and more expensive), and where a single parastatal controlled all aspects of the chain from input supply to lint marketing before liberalization, the orderly privatization of the parastatals led to effective duopolies. In 1995/96, the Cotton Company of Zimbabwe Ltd (Cottco), the privatized successor to CMB, was joined in the market by Cargill. Cargill established a 25 percent market share in its first two years of operation, a share that has remained stable ever since. In Zambia, the assets of Lintco were sold in two parts to Lonrho and Clark Cotton. These operations were subsequently sold to Dunavant and Cargill, respectively, and these companies still dominate the market.

In the first few years after liberalization, the concentrated sectors were found to perform best (Poulton et al. 2004). Zimbabwe completed its transition to a smallholder-based system, with

Cottco's credit scheme (based on the scheme established by CMB in 1992) an important part of the story. Strict attention to quality by Cottco and Cargill allowed the sector to maintain its excellent reputation for quality on international markets. Meanwhile, Zambia achieved a strong increase in production because of a gradual increase in yields among established farmers and large increases in the number of farmers. This production increase was temporarily interrupted by side selling of seed cotton, caused by the entry of a number of new firms in 1998 and 1999. However, several of these firms exited when world lint prices fell. Meanwhile, Dunavant, which had bought out Lonrho during this period, introduced a system of independent "distributors" to handle credit and extension provision to farmers, which contributed to a further expansion of production. Clark's more traditional system of extension agents was quite effective—the distributors trained by Dunavant focus primarily on input distribution and credit recovery, and only to a secondary degree on extension advice.[37] In addition, both companies spearheaded campaigns against polypropylene contamination, which laid the foundation for a sizeable increase in the price premium that Zambian lint now receives on world markets.

Meanwhile, Tanzania and Uganda struggled to support farmers in a highly competitive output market. Efforts by individual companies to provide input credit were quickly abandoned because the credit could not be recovered. The Uganda Cotton Ginners and Exporters Association experimented with an innovative scheme to provide chemicals to producers on a sector-wide basis, but had to abandon the effort for various reasons. Eventually, in 2003 the sector moved to a zoning system that severely restricted competition, as a way to give ginneries the security to invest in extension provision and input supply.[38]

Starting in 1999, Tanzania began experimenting with an input trust fund to provide farmers with minimal access to chemical input. This fund was subsequently replaced by a passbook (forced saving) system, which was superior in a number of ways. As in Uganda, a sector-wide solution had to be sought to the input supply challenge because the private incentives do not exist for individual companies to provide input in a highly competitive output market. An additional challenge in Tanzania was maintaining the quality of seed cotton and lint. Liberalization accelerated the decline in lint quality that had begun earlier, because seed varieties were soon mixed and a scramble for seed cotton undermined farmers' incentives to supply good quality seed cotton to buyers.

After a short-lived boom induced by high world prices, production fell sharply in Tanzania after liberalization. In Uganda, it has remained disappointingly stable since liberalization. The challenges of increasing productivity and production in a sector with numerous small to medium ginners have encouraged multistakeholder collaboration in both countries, but with uneven results. In Tanzania, at least, this collaboration now appears to be bearing some fruit.

Meanwhile, a fairly dramatic change in sector organization has occurred in Zimbabwe since 2001 and a similar change may now be occurring in Zambia. In Zimbabwe, the onset of economic crisis in 2001 made acquisition of foreign exchange a top priority, and cotton production appeared an attractive way of achieving this goal. In addition, the real exchange rate depreciated spectacularly during 2001 and 2002, but the existing cotton companies did not pass on the benefits to farmers. As a result, the total number of ginners rose from 5 in 2000/01 to 17 in 2006/07. The overall effect of this dramatic change is still to be determined, but it is clear that the sector now faces similar challenges on quality control, input supply, and extension provision as described above for Tanzania.

Established players in the sector realized that the new circumstances required a new regulatory framework. In 2004 cotton sector stakeholders presented a draft set of regulations to the Zimbabwe Minister of Agriculture, but as of mid-2008 the changes had not been approved. Instead, consensus was reached on stricter licensing procedures for 2007 that required all cotton lint exporters to demonstrate that they had supported smallholder cotton farmers. As in Uganda and Mozambique, it appears that attempts to strengthen incentives for provision of preharvest services by ginning companies will come at the cost of reduced competition in the seed cotton market.

In Zambia, new companies have also entered the sector since about 2005 and there has been a resurgence of side selling of seed cotton. Unlike in 1997, however, these new entrants are backed by investment in new ginneries, so it seems unlikely that the sector will return to its former duopoly structure. As in Zimbabwe, the new entry appears to be associated with low seed cotton prices since the early 2000s, and also to instability in the real value of the exchange rate. In Zambia, the kwacha appreciated rapidly (but temporarily) before the 2006 election, limiting the prices that companies could pay for seed cotton. This, however, occurred at a time when farmers were already dissatisfied with prices, compounding their dissatisfaction and making them willing to switch allegiance to new players.

In Zambia there has been an intermittent debate about a new regulatory framework for the sector ever since the first burst of new entry and side selling in 1997. The major points of contention have been enforcement of contracts and prompt resolution of disputes when they occur. There have been suggestions of establishing fast-track courts for this purpose, and of amending the Agricultural Credits Act. However, the main stakeholder focus has been on ensuring passage of a revised Cotton Act; as of August 2007, the Minister of Agriculture had submitted the revised act to Parliament, but it has not yet been passed (Tschirley and Kabwe 2007a).

Regulated sectors include national monopolies and local monopolies. In these systems, cotton ginning company(ies) have an exclusive right—and a non explicit obligation—to buy all cotton seed offered by farmers either over the whole territory of the country (national monopoly) or over a delimited geographic area (local monopoly). They feature in general a single-channel marketing system for both inputs and outputs. Monopsony would be the right economic denomination, but, since there is generally a mirror image between the structure of the seed cotton market, lint sales and input supply, the term monopoly is in effect commonly used as a shorthand. Local monopolies could alternatively be designated as zonal or sub-national monopolies. However, the term local monopoly will be used throughout the rest of this study.

Overall, therefore, the story in ESA is one of heterogeneity in sector structure postliberalization, of ongoing evolution in those structures, and new challenges as to how the different cotton sectors should be regulated. While WCA does not have the same clear experience of "liberalization" as ESA, the two regions do have plenty in common in their ongoing search for effective responses to changing global conditions and shared challenges.

# Chapter 4: A Typology of African Cotton Sectors
## Colin Poulton and David Tschirley

For the purpose of this analysis, five types of African cotton sectors are delineated, based on the structure of the market for seed cotton purchase and the regulatory framework in which firms operate. Both of these factors influence firm conduct, which influences sector performance. The types are set out in figure 4.1 in a decision tree framework. The first distinction is between "market-based" and "regulated" sectors. Because all markets function within some type of regulatory framework, regulated in this context means a sector in which competition for the purchase of seed cotton is not allowed. This type of regulation was standard throughout African cotton sectors before the early 1990s, and has continued in most of Western and Central Africa (WCA) to the present.

Regulated sectors include national monopolies and local monopolies. In these systems, cotton ginning company(ies) have an exclusive right – and a non explicit obligation – to buy all cotton seed offered by farmers either over the whole territory of the country (national monopoly) or over a delimited geographic area (local monopoly). They feature in general a single-channel marketing system for both inputs and outputs. Monopsony would be the right economic

**Figure 4.1 Decision Tree for Cotton Sector Typology**

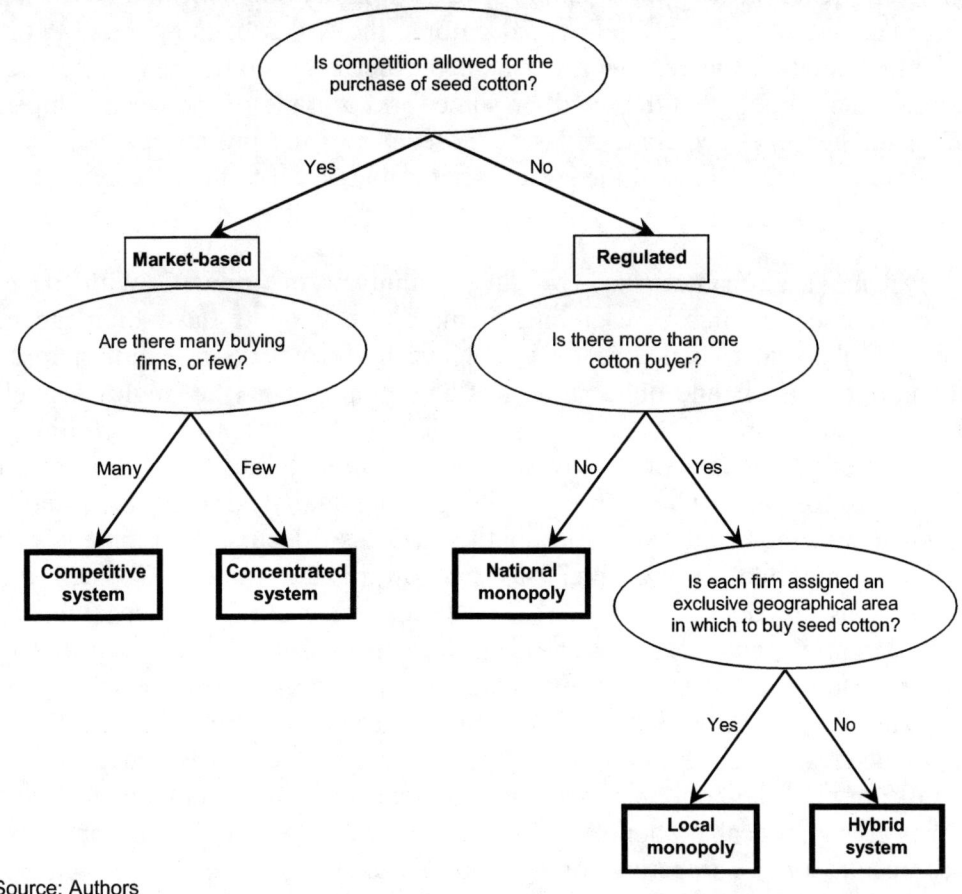

Source: Authors

35

denomination, but, since there is generally a mirror image between the structure of the seed cotton market, lint sales and input supply, the term monopoly is in effect commonly used as a shorthand. Local monopolies could alternatively be designated as zonal or sub-national monopolies. However, the term local monopoly will be used throughout the rest of this study.

For market-based systems, two further distinctions can be made: those with "many" buyers of seed cotton (competitive systems) and those with "few" such buyers (concentrated systems). Necessarily subjective, this distinction is nevertheless meaningful when one compares a country such as Tanzania (more than 30 buyers) with Zambia before 2006 (one dominant buyer, one large competitor, and two or three other very small buyers).

## Competition and Coordination

Poulton et al. (2004, 521) defined coordination as "effort or measures designed to make players within a market system act in a common or complementary way or towards a common goal." They noted that the pursuit of effective coordination "may…require effort or measures designed to prevent players from pursuing contrary paths or goals." In the neoclassical ideal of perfect competition, the only coordination required is vertical coordination between players at different levels of the system, and this coordination is fully achieved through the price mechanism. Coordination among players at one level in the system—horizontal coordination—does not appear in this model. Yet North (1990) argued that implicit in the perfectly competitive model, and essential to any real world approximation of it, is a highly sophisticated set of institutions that make information available and define and enforce the "rules of the game." Poulton et al. (2004, 521) suggest that in the real world, where the perfectly competitive ideal never fully holds, it "becomes more likely that there will be some form of trade-off between competition and coordination." This hypothesized trade-off is at the center of the proposed typology, and it is possible to summarize expectations of the sector types with respect to their likely performance in each dimension.

Competitive systems are characterized by large numbers of ginners, with open market competition for seed cotton purchase among them. These systems have high incentives for efficiency, but are likely to find it difficult to achieve horizontal coordination across firms to ensure input credit, extension, and lint quality. National monopolies (the single-channel systems that have been common in WCA) solve the coordination problem by consolidating most downstream activities in a single firm. However, this solution comes at the cost of potentially very low incentives for efficiency; for example, ginning and overhead costs may rise and performance on input credit and extension quality may also decline over time. Concentrated systems and local monopolies are likely to lie toward the middle in each dimension. Concentrated systems are dominated by two or perhaps three major ginners, which compete directly for the right to buy seed cotton from farmers (that is, there is no geographical segregation of their activities). The competition is focused on getting producers to sign up with a particular company for a given season and tends to be based as much on the quality of services provided to producers as on seed cotton price. Once a growing season is under way, the major ginners generally respect the contracts that each has reached with particular producers for the duration of that season. Local monopolies do not rely on this self-policing approach, instead prohibiting companies from competing for seed cotton outside their specified zones. Expected differences in performance between concentrated systems and local monopolies are not large;

**Figure 4.2 African Cotton Sector Typology**

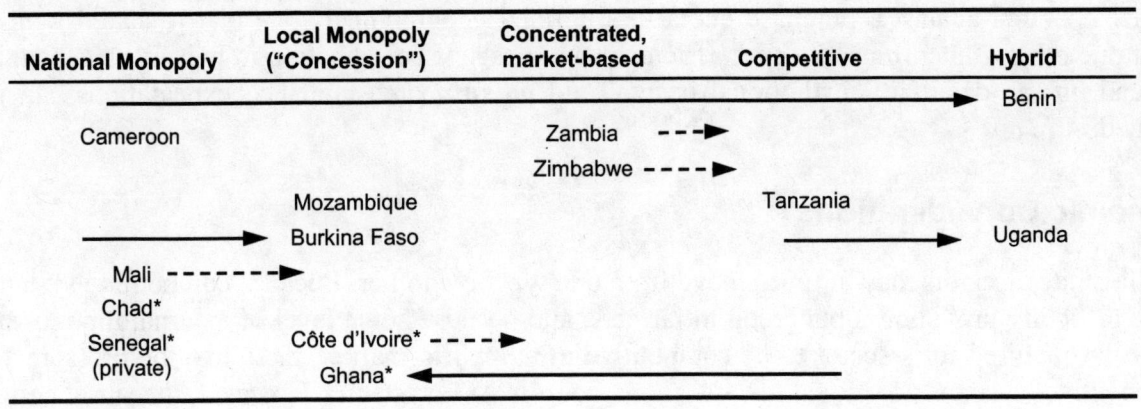

Source: Authors.
Note: Dashed arrows mean planned or still in progress. Solid arrows mean the evolution has been completed.
* Not included in this study.

however, policy, regulatory capacity, history, and other factors can heavily affect firm behavior and performance.

According to figure 4.1, the basic difference between competitive and concentrated systems lies in the number of ginning firms within the sector. Thus, the concentrated sectors shown in figure 4.2 (Zambia and Zimbabwe) were both, until fairly recently, effectively duopolies in which the top two firms accounted for 90 percent or more of seed cotton purchases. By contrast, in Tanzania there are about 30 ginners, the top five of which only account for about 40 percent of seed cotton purchases (and these top five typically change from year to year). However, the number of ginning firms within the sector is essentially a continuum, so the dividing line between the two systems is not entirely clear cut.

However, as noted above, there are also important differences in the nature of competition within competitive and concentrated systems. This came into sharp focus in Zimbabwe after 2001, where, as new firms entered the sector, the clash of competitive cultures arguably caused more problems for established firms than the fall in the concentration ratio. During this period the share of seed cotton purchases accounted for by the top two firms has fallen to between 70 and 80 percent, which is at least 10 percentage points less than before 2001, but is still heavily concentrated. However, the conduct of many of the new firms has resembled that of firms in the competitive Tanzanian sector (offering few preharvest services and willing to compete on price at harvest time, thereby encouraging side selling by farmers who have received credit from established companies) rather than that of the two established firms within the sector. The Zimbabwe experience shows that it is difficult for the two competitive cultures (and the visions for the cotton sector that underlie them) to coexist within a single sector.

Finally, hybrid systems are a potentially diverse group, emerging either out of attempts to liberalize a national monopoly (Benin) or to solve the problems unleashed by liberalization in a sector with a competitive structure (Uganda). Thus, Benin has fewer than 10 ginners, each with a purchasing quota but no fixed geographical zone in which to make those purchases. The sector also has a complex clearinghouse approach to the provision of input and purchase of seed cotton. Uganda has about 30 ginning firms that initially competed against each other after reform. From about 2002 through 2007, in response to input credit problems created by that competition, each ginner operated against a purchase quota in a defined geographical zone that it shared with at least one other ginner. Incentives for efficiency and costs of coordination in hybrid systems thus depend on the details of institutional design, and no firm preestablished expectations can be formed.

## Dynamic Considerations

As illustrated, sectors may at times move from one type to another. Because competitive systems have difficulty providing input credit to farmers, and because some level of external input use is typically required for a sector to be competitive in the world market, there may be pressure for competitive systems to move toward more coordinated systems. However, because cotton ginning does not have large economies of scale (especially where roller gins are used), a move from a competitive structure toward an unregulated, concentrated system is unlikely. Therefore, if competitive systems change, it is likely to be toward one of the regulated types. Because it is impractical to impose a single monopoly on a system with many private firms, this movement is very likely to be to a local monopoly or hybrid system.

In the study's sample of countries, Tanzania and Uganda both have competitive structures that undermined input credit after reform. Yet they have followed very different paths in dealing with this challenge. Tanzania has maintained its commitment to a competitive system while experimenting with innovative approaches to partially address the input credit problem. Uganda moved to a hybrid system that keeps all ginners operating, but attempts to eliminate all direct competition among them for seed cotton, while investing heavily in training of farmers. Box 4.1 suggests possible explanations for these radically different choices in two countries with very similar prereform histories and nearly identical post-reform competitive structures.

Management theory suggests that national monopolies are likely to show growing inefficiency over time.[39] This inefficiency can eventually undermine performance on input credit, extension, and cost competitiveness of the enterprise. If these systems change, the direction of change depends on policy choice. In principle, a single-channel system can move quickly to a competitive system if free entry is allowed and competition is not regulated. In practice, the cultural norms that resulted in the single-channel system will often (though not always) lead to a more deliberate reform, either toward a local monopoly system (Burkina Faso and Mali) or to a concentrated, market-based system (Zambia and Zimbabwe).

In Mali, the move to a local monopoly system has been proposed, but has yet to be implemented. The changes in Zambia are recent and deserve to be examined further. Uganda has a large number of private ginning companies that competed heavily at one point. The sector has experimented with two different hybrid regulatory approaches over since the mid- to late 1990s to solve its input credit problem.[40] With regard to firm conduct, we expect that Uganda and Benin both now resemble a local monopoly system.

Figure 4.2 maps the study sectors (and a few additional examples) onto the typology, along with indications of how sectoral structures have changed, if at all, since the onset of sector reform on the continent in 1994. Solid lines depict definitive changes, while dashed lines suggest changes that may be under way.

**Box 4.1 Is This Typology Specific to Africa?**

The typology presented in this chapter grew out of work in East and Southern Africa (ESA), and needed only minor modifications to usefully incorporate WCA. Two questions thus emerge. First, are the sector types identified here found in other areas of the world, and does an assessment of their strengths and weaknesses stretch beyond African borders? Second, does the rest of the world exhibit sector types not found in the typology, and do these provide any glimpses into the likely future evolution of African systems? More to the point, do other sector types provide clues about what types of change policy makers and stakeholders should be encouraging in their sectors?

Characterizing cotton sectors worldwide is quite complex and would require further investigation. Yet a number of trends can be observed in the institutional evolution of cotton sectors throughout the world:

- Cotton ginning is not by nature a heavy industry. Hence, there are numerous examples of farmers grouped in associations, as well as farm-based agribusinesses, that are engaged in cotton production and ginning in some of the major exporting countries, such as Australia, Brazil, and the United States. There are similar cases in Africa, such as SICOSA in Côte d'Ivoire, which is the ginning company established by a cotton farmer union (URECOS-CI) at the end of the 1990s. In other cases, ginning is a service provided on a fee basis (toll ginning) to farmers, who retain ownership of the final products (lint and seeds).

- There are few remaining examples of state-owned enterprises buying raw cotton from farmers through a single-channel (national monopoly) system. Even former Soviet republics in Central Asia, which are large exporters of raw cotton, have, with the exception of Turkmenistan, moved away from the national monopoly system to various degrees. These countries include Kazakhstan, Tajikistan, and Uzbekistan.

- Local monopolies seem to be transitional arrangements in the evolution from a single-channel system rather than a permanent sustainable organizational model. Most cotton sectors in the world can be characterized as either concentrated or competitive.

- Cotton production and ginning activities are very seldom integrated with downstream industries such as spinning, weaving, and textile manufacture, except in some particular locally favorable conditions (China, India, and Turkey).

It can be inferred from this short review that change in African cotton sectors is moving in similar directions to what can be observed among other major cotton producers in the world: retreat of governments and state cooperatives from industrial and commercial activities, growing empowerment of farmer groups in the management of the cotton sectors and in ginning and exporting activities, and sharply increased investment in local cotton industries by international cotton merchants and commodity-based multinationals when they see opportunities at the country level. However, this is not to say that convergence in modes of sector organization is imminent. The organization of cotton sectors in Africa faces specific challenges as a result of two factors: (i) the high input intensity of cotton production, and (ii) the weakness of markets for input and—arguably even more important, given the lack of capital of most African smallholder households—seasonal finance in Africa. As long as the seasonal finance constraint remains, the issue of the optimal form of cotton sector organization in Africa will continue to be complicated and convergence toward forms of cotton sector organization observed in other parts of the world will remain partial.

## Predicted Strengths and Weaknesses of Different Sector Types

Poulton et al. (2004) identified four main challenges facing smallholder-based African cotton sectors:

- provision of input credit to farmers

- maintenance of quality control

- maintenance of a high quality research system and effective extension of resulting research knowledge and products

- payment of an attractive seed cotton price

Table 4.1 summarizes some of the strengths and weaknesses hypothesized about different sector types. In the absence of a strong, high capacity government regulatory agency, Poulton et al.

**Table 4.1 Trading Off: Strengths and Weaknesses of Different Sectoral Types**

| Characteristic | National monopoly | Local monopoly | Competitive | Concentrated |
|---|---|---|---|---|
| Nature of competition among ginners | None | Concession rules may create competition; some emulation across zones on costs, prices, and services | High, tends to focus on seed cotton pricing | Moderate, as much on service provision as on pricing; price leadership often observed |
| Potential strengths | Conditions may be conducive to provision of input credit, quality control, extension, and research | Conditions may be conducive to provision of input credit, quality control, extension, and research | Seed cotton pricing | Conditions may be conducive to provision of input credit, quality control, extension, and research |
| Potential weaknesses and major challenges | Cost control; maintaining attractive producer prices; political interference | In presence of a weak state, requires strong farmer organizations to ensure setting and implementation of transparent rules for concession allocation and performance evaluation | Preharvest service delivery; quality control; accountability of government agencies responsible for these functions | Seed cotton pricing heavily dependent on internal objectives of dominant companies; instability of market structure |

*Source:* Authors.

(2004) hypothesized that coordinated sectors (national and local monopoly, concentrated) will be more likely to respond effectively to the first three of these challenges, whereas competitive sectors will be more likely to generate attractive seed cotton prices for farmers. These hypotheses are further explored in this book.

While these hypotheses assume the absence of a strong, high capacity government regulatory agency, the earlier summary of historical experience in WCA and ESA serves to emphasize the importance of sector governance to good performance in all five sector types. Concentrated sectors can perform well with minimal input from the state. Good performance then depends on private coordination among the dominant ginning firms. However, farmers are reliant on these companies' continued ambitions for expansion for an attractive seed cotton price. As has been seen with new entry in both Zimbabwe and Zambia in recent years, such sectors can be contestable, which should provide an incentive to incumbent firms to keep paying attractive prices to producers. However, if they cease to continue paying attractive prices and other firms enter, the change in sectoral structure brings challenges as well as advantages. So far, there are no examples of market forces alone producing significant concentration within a cotton sector from a more competitive base. Thus, the change in sectoral structure from concentrated to competitive is likely to require a new approach to regulation and a more active role for the state.

Failures in the market for seasonal finance (and hence input access) and quality control mean that competitive sectors have had to look to government agencies to play an active role that goes beyond the provision of conventional public goods. This expanded role raises the potential for government failure—and indeed, government agencies in both Tanzania and Uganda (classified here as a hybrid system but with a large number of private ginning companies) have faced such challenges. Thus, the development of mechanisms by which other stakeholders can hold government agencies accountable for their actions becomes critical to overall sector performance.

The regulatory challenge is arguably greatest in the local monopoly system, where theory predicts that performance will be enhanced by the setting and impartial implementation of transparent rules for concession allocation, periodic performance evaluation, and reallocation. This is a tall order even in a developed economy, let alone an African economy with much less experience at government capacity building. However, there may be strong pressures for a local monopoly system from cotton companies that are skeptical of their ability to make a market-based system work. In addition, the fact that a local monopoly system has a legal foundation may give it a degree of stability.

A range of variants is possible on the basic local monopoly model. Specifically, decisions about pricing may be made at a central level (through some sector-wide price-setting mechanism, as currently happens in Burkina Faso and Mozambique) or a decentralized, that is, concession, level. This may also be true of other decisions. As with hybrid systems, the performance of a local monopoly is likely to be heavily influenced by the detailed rules of the game governing such decision making.

The past history of national monopoly systems suggests that one of the biggest challenges is how to prevent politicians from meddling in sector governance. Cameroon shows that this can be accomplished (albeit perhaps in special circumstances), while experiences in Burkina Faso and Tanzania show that this challenge is not confined to national monopoly systems. One of the main justifications for intervention by politicians is to ensure that producers receive a fair price for their seed cotton. However, history is replete with cases where political intervention achieved the opposite outcome (a recent example being Mali from 1994 to 2002). Another key challenge facing national monopoly systems, therefore, is how to ensure that farmers' interests are safeguarded, in particular that company operating costs are kept under control so that attractive prices can be paid to producers.

## Conceptualizing the Links between Cotton Sector Organization and Performance

The previous section set out a number of expected strengths and weaknesses associated with different forms of cotton sector organization. In this section the key links between cotton sector organization and various dimensions of performance are discussed more systematically. The section culminates in discussion of a series of indicators by which cotton sector performance will be assessed within this book. Figure 4.3 presents the conceptual framework that underlies much of the subsequent analysis in the book.

Given the various links in the chain, one can think of cotton sector performance at various levels. First, there are the processes and services that are under the direct control of cotton companies and other stakeholders within the cotton sector. These include public goods generation within the sector, the services delivered to cotton farmers, seed cotton pricing decisions, and the mechanisms that companies and other stakeholders put in place for controlling and enhancing cotton quality. These are shown in the rectangular boxes near the top of figure 4.3. Performance at this level focuses primarily on the quality of services provided to farmers and the prices paid. The process indicators for assessing performance at this level are analyzed in section III of the book.

At the company level, the extent and efficiency of these processes and services influence the overall cost of operations. More extensive preharvest services raise company costs, but so too does a paying a higher price for seed cotton. Cost of operations is also heavily influenced by the efficiency of ginning, which, theory suggests, is likely to be closely linked to cotton sector organization. Cost of operations may be thought of as an intermediate performance indicator, but one that has a major influence over both company profitability and the competitiveness of the sector in final markets.

Meanwhile, at the farm level, the public goods generated and the services delivered by cotton companies are major determinants of seed cotton yields (although exogenous factors such as soil fertility also come into play—portrayed with dotted ovals in figure 4.3).[41] As with cost of company operations, seed cotton yields may be thought of as an intermediate performance indicator. Along with seed cotton prices, yields are a major determinant of farm-level profitability of cotton production. At this point, an important feedback loop comes into play, whereby high profits and cash incomes from cotton can facilitate the acquisition of capital assets, such as draft power and related equipment, that can further enhance the scale and efficiency of cotton production and of the whole farm enterprise.

Section IV of this book analyzes performance at both company and farm levels, focusing on outcome indicators. At the farm level, yields and returns to farmers are examined. At the company level, the focus is on cost efficiency and then on overall competitiveness in the lint market, which is a function of the costs of production and the price that a sector realizes for its lint.

Ultimately, the goal would be to link organization of a cotton sector to its contribution to poverty reduction. However, many exogenous factors mediate the impacts of cotton on poverty reduction within an economy and modeling the impact on poverty within the focus countries is beyond the scope of this book. The analysis is therefore restricted to outcome indicators because there are indications of clear relationships between the indicators chosen in this book and the impact of the cotton sector on poverty within a given country.

Beginning at the farm level, increasing profitability of cotton production for farmer households does not guarantee poverty reduction, but should contribute to it. Directly, the incomes of cotton farmers (many of who may start poor) increase. Indirectly, higher profitability of cotton production should encourage more hiring of labor, while higher incomes should set off consumption multiplier effects within the local economy. As profitability of cotton production increases, it may also encourage more households to engage in cotton production. The impact on household incomes may often be quite small, unless the cotton sector is able to sustain a higher rate of productivity growth over time relative to competing crops and activities.

# Figure 4.3 Linking Cotton Sector Organization and Performance

*Source:* Authors.

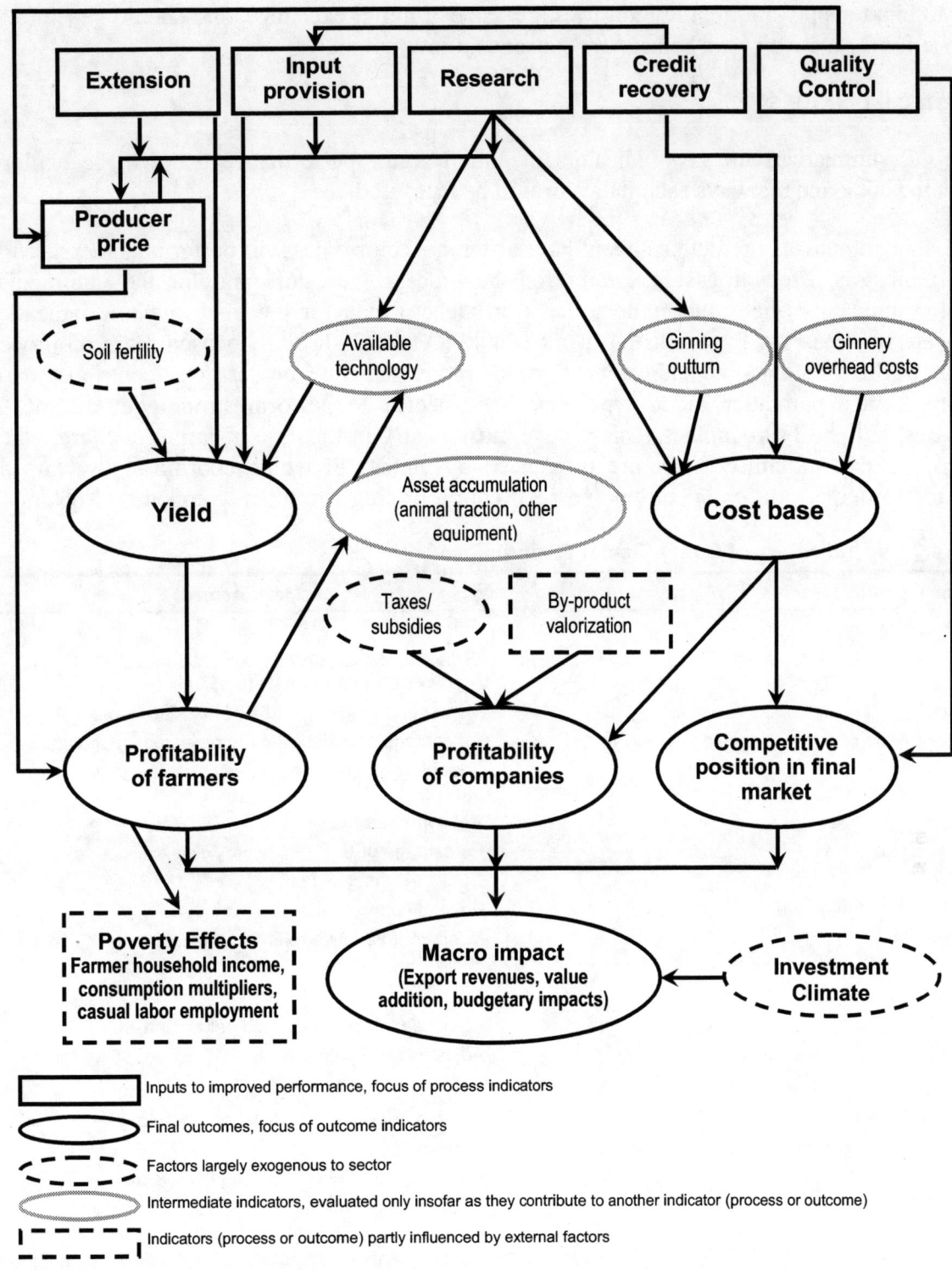

*Source:* Authors.

At the macro level, a healthy cotton sector should also contribute to economic growth, and thereby to poverty reduction. It creates value within the economy, has backward and forward links to input supply and textile industries, generates foreign exchange, and should generate tax revenue for the state that can be used for subsequent investment.

## SELECTED INDICATORS

Table 4.2 summarizes the key indicators of cotton sector performance that will be explored within the book and the ways each indicator will be measured.

Table 4.3 summarizes predictions about how different sector types will perform as measured by these indicators. Predictions are clearer for the process indicators and for the intermediate outcome indicators (yield and company cost efficiency) than for the final outcome indicators. There are two reasons: First, existing work (Poulton et al. 2004) has focused more on process indicators than on outcome indicators. Second, more than one process can contribute to an outcome and a particular sector type may be expected to perform strongly in one of the processes, but poorly in another. This is seen most clearly in the case of farmer welfare, where returns per day of family labor are influenced by yields (in which coordinated sectors are expected to outperform competitive) and seed cotton pricing (in which competitive sectors are

**Table 4.2 Key Indicators of Cotton Sector Performance**

| Type of indicator | Measurement |
|---|---|
| **Process indicators** | |
| Quality and marketing | Estimated average realized premium over the Cotlook A Index on world markets (US$/lb lint) |
| Pricing | Mean percentage of FOT price paid to farmers |
| Input provision | Percentage of cotton farmers receiving input credit |
| | Adequacy and quality of input credit package, if provided |
| | Repayment rate |
| Extension | Percentage of companies providing assistance |
| | Qualitative assessment |
| Valorization of by-products | Price of cotton seeds |
| Research | Number of varieties released and taken up in past 10 years |
| **Intermediate outcome indicators** | |
| Yield | Kg of seed cotton produced per hectare |
| Company cost efficiency | Adjusted farm gate price to FOT cost (US$/kg lint) |
| **Final outcome indicators** | |
| Farmer welfare | Returns per day of family labor (US$/day) |
| Number of farm households participating in sector | Number |
| Overall competitiveness | Ratio of total FOT cost to total FOT value |
| Macro impact | Total value added per capita (including value of seed sales) |
| | Net budgetary contribution per capita (taxes paid minus transfers received) |

*Source:* Authors.
*Note:* FOT = Free on truck.

# Table 4.3 Expected Performance along Key Indicators

| | National monopoly Local monopoly | | Concentrated | Competitive |
|---|:---:|:---:|:---:|:---:|
| **Process indicators** | | | | |
| Quality and marketing | Medium | | High | Low |
| Pricing | Low if left to companies alone | | Low | High |
| Input provision | High | | Medium | Low |
| Extension | High | | Medium | Low |
| Research | High | | Medium | Low |
| Valorization of by-products | No clear prediction | | | |
| **Intermediate outcome indicators** | | | | |
| Yield | High | | High | Low |
| Company cost Efficiency | Low | | Medium | High |
| **Final outcome indicators** | | | | |
| Farmer welfare | No clear prediction | | | |
| Overall competitiveness | No clear prediction | | | |
| Macro Impact | No clear prediction | | | |

*Source:* Authors.

expected to outperform coordinated). In table 4.3, therefore, "no clear prediction" means that this is an empirical question worthy of additional research.

This study suggests that performance on input credit and extension is likely to be correlated across sector types: sectors performing well compared with other sector types on input credit are likely to perform well on extension. This correlation is in large measure due to the complementarity of the two activities, which stems from two sources. First, agents delivering input need to transmit at least some minimum level of knowledge of how to use the input so that the company recoups reasonable value (in the form of increased production) from the investment. Second, extension work can "free ride" on many of the fixed costs (such as travel cost and time) associated with input delivery. Yet delivery of input and extension advice are not likely to be perfect complements. Within any sector type, this suggests that performance on extension is likely to be weaker, or more fragile in the face of stress, than performance on input credit, for three reasons. First, delivering anything other than the most basic extension advice directly linked to input use requires a higher level of training of field agents than delivery of input. Second, extension delivery, and especially its effectiveness, is more difficult for companies to monitor than is input delivery. Third, even for companies committed to increasing farm-level productivity, the return to extension is likely to take longer to appear and to be more difficult to identify than the return to input delivery. For all these reasons, commitment to extension is likely to be harder to maintain than commitment to input delivery; especially where systems come under stress (for example, because of increased competition in concentrated systems), extension effort may be sacrificed for narrower loan monitoring and recovery.

In closing this chapter, it is important to note that the type and quality of sector regulation will have a strong effect on performance for all indicators. The job of regulation may be thought of as seeking correction in areas where an unregulated sector is likely to perform poorly, while preserving that sector's strengths. The corollary is that the predictions in table 4.3 may be most accurate in situations where regulation is weak, which it typically is in Africa. However, as has already been described (chapter 3), there are ongoing efforts to build more effective regulatory regimes for the cotton sector in several of the study's focus countries.

## Box 4.2 Factors in Addition to Structure That Influence Behavior and Performance

The sectoral typology developed in this book focuses heavily on the structure of the market for seed cotton. This structure is seen as a key determinant of the types of challenges that a sector will have most difficulty meeting, and thus of the types of regulatory structures that are needed to safeguard good performance and improve poor performance. For example, the typology suggests that highly competitive sectors will be unable to provide input credit and extension or to safeguard quality, but will be likely to pay attractive prices; concentrated sectors are predicted to perform well on input credit, extension, and quality, but to pay less attractive prices. Comparing Uganda with Tanzania, and Zimbabwe with Zambia, highlights the usefulness of the typology while making plain that structure is not the only factor that influences behavior and performance. In each case, the structure that emerged out of reform—highly competitive in Uganda and Tanzania, very concentrated in Zambia and Zimbabwe—did determine which challenges (input credit, extension, productivity growth, seed cotton pricing, lint quality, ginning efficiency) countries had most difficulty in meeting. Yet in each case, variations in history, management, or geography have led to important differences in behavior and performance.

These differences are most stark in Uganda and Tanzania, despite remarkably similar structures and histories before and immediately after reform. In both countries, the prereform cooperative-based systems led quickly after reform to markets with 20 to 30 buyers competing intensively on price for farmer production. Farm prices improved, but input supply and extension collapsed. Initial efforts to solve the problem in both countries involved moving input supply to the public sector to allow competition among companies in the output market. Both efforts achieved some success but failed after two years because of management and design problems.

Since the failure of these initial attempts, the two countries have moved in dramatically different directions (chapter 6). Uganda reversed its previous course, using a zonal quota system to eliminate competition in the output market in an effort to facilitate coordination by ginners on input supply and extension. Tanzania, meanwhile, maintained its competitive output market and abandoned public sector input distribution. Instead, stakeholders in Tanzania used innovative approaches, such as the so-called passbook system, that require close cooperation between public and private sectors to provide seed and some minimal level of insecticides to farmers.

Several factors help explain why Tanzania and Uganda have chosen such different paths. First, the geographical scope for expanding cotton production under a low-input approach is greater in Tanzania than in Uganda, at least as long as the far north of Uganda remains insecure. For now, ginners in Uganda must try to increase production in relatively small areas already under production. If this line of reasoning is correct, once the northern region becomes secure, incentives for ginners to operate the quota system may decrease.[a] Second, farmers in Uganda may have more remunerative cash cropping options than in Tanzania. Providing some substantial input credit or extension, or both, is thus imperative if ginners are to attract farmers to cotton; again, this may not be the case in the more remote north of the country. Finally, greater cultural homogeneity among Ugandan ginners may have facilitated collective action approaches that were infeasible in Tanzania.

Differences in Zambia and Zimbabwe have less to do with regulatory approaches—both are attempting to consolidate their concentrated structures—than with the productivity that each has achieved. Zimbabwe's annual mean yield from 1995 through 2004 was about 825 kg seed cotton per hectare (ha), compared to about 625 kg/ha in Zambia (chapter 10). Even in 2006, after several years of macroeconomic crisis and disrupted input distribution resulting from the entry of new players, the best-yielding 20 percent of Zimbabwean farmers achieved mean yields of 1,750 kg/ha, while the top 4 percent of Zambian farmers reached only about 1,200 kg/ha. Zimbabwe's superior performance in this regard is driven by the continuing influence of past performance. Before reform in 1994, Zambia's Lintco was a small and declining company, and little if any cotton research took place in the country. In contrast, Zimbabwe developed its Cotton Marketing Board (CMB) and an effective research system in the 1960s and 1970s to serve commercial farmers. After Zimbabwe's independence in 1980, the CMB built systems for effective input credit and extension to new smallholder farmers; farmers not included in these systems were able to self-provision at least a basic set of input in the country's relatively developed (by Sub-Saharan African standards) private input markets. Cottco, the private sector successor to CMB, continued this effective performance into at least the early 2000s. Even today, after temporarily scaling back its input credit support in 2005, the company provides high quality support to many farmers.

a. Indeed, the north of Uganda is beginning to open as this book is being finalized (mid-2008), and Uganda abandoned the quota system for the 2007/08 production season. It is not yet clear whether this decision was driven in part by the prospect of expansion into northern areas of the country.

This book considers the cotton sector as a whole, including the broad range of activities from seed cotton production through ginning to marketing of the resulting products (lint, seeds, oil, and cake).[42] As a result, the analysis includes a description of the way cottonseed oil industries are organized and how they contribute to the sector's performance. An argument made in this study is that performance of oil and cake sectors is becoming increasingly important and thus requires more attention than it has received to date. Also, in some cotton sectors outside Africa, farmers own and retain their cotton seeds after ginning, meaning that the performance of these markets directly affects farmers.

Table 4.2 features the valorization of seed cotton by-products as a process indicator. Similarly, figure 4.1 presents it as an endogenous variable, that is, something directly affected by the organization of the cotton sector. In fact, while the organization (and hence performance) of oil and cake markets are related to the organization of the cotton sector, the nature of this relationship is not the same as that between sector organization and performance on, say, input supply or quality control. Historically, in WCA, the organization of the oil sector was bound up with the national monopoly in cotton ginning—all part of the overall development approach to the cotton filière (supply chain).[43] While this is history more than an inherent part of a national monopoly in cotton ginning, it has determined the structure and evolution of the cottonseed oil industries in most of the WCA countries. Meanwhile, market-based cotton sectors are generally associated with more or less competitive oil markets.

# SECTION III
# COMPARATIVE ANALYSIS: CORE ACTIVITIES AND SERVICE DELIVERY

# Chapter 5: Pricing Systems and Prices Paid To Growers
### John Baffes, David Tschirley, and Nicolas Gergely

Until recently, most cotton sectors in West and Central Africa (WCA) were national monopolies, requiring an administered approach to farmer price setting. While details vary across countries, each country in WCA followed broadly similar approaches to this issue. Since the onset of reform in East and Southern Africa (ESA), cotton sectors have included all sector types described in chapter 4 except national monopolies: local monopolies (Mozambique), concentrated (Zambia and Zimbabwe), competitive (Tanzania, Uganda until the early 2000s), and hybrid (Uganda since 2003). Predictably, approaches to price setting in ESA have been as diverse as the countries' sectoral and regulatory structures.

The summary indicator for pricing performance is the percentage of the free on truck (FOT; ex ginnery) lint price paid to farmers for their seed cotton. The typology laid out in the previous chapter generates clear expectations of performance on this measure for market-based systems: competitive systems should pay the highest FOT share, while concentrated are likely to pay the lowest. Performance in national and local monopolies is not fully predictable from the typology because political factors and the strength of farmer associations weigh more heavily in these sectors. Before presenting summary results, pricing mechanisms in the two regions are briefly reviewed.

## Pricing Mechanisms in WCA

Pricing mechanisms in Francophone WCA have a remarkably similar historical background across all countries, based on a commonly accepted principle that single-channel systems require fixed prices: unique for the entire cotton-growing area in the country (panterritorial), fixed throughout the cropping season (panseasonal), and announced publicly before sowing. Another major feature is the guarantee of purchase by the cotton company of all quantities of seed cotton offered at the official price. Typically, during the 1970s and most of the 1980s, the Ministry of Agriculture announced the producer price for seed cotton before the planting season, and the cotton company was mandated to purchase the raw cotton from farmers throughout the country. There was intense bargaining each year between the government and the cotton company based on standard costs for input and services including ginning (*barêmes*) because the cotton company was assuming the financial risks of the guaranteed price and tried to cover risks and overhead through well-negotiated *barêmes*. This bargaining process, which featured little transparency, was replaced at the end of the 1980s by performance contracts negotiated and signed between the government and the cotton companies that set performance targets and costs. As noted in chapter 3, the results of the performance contract approach were generally disappointing.

Since the successive crises of the early to mid-2000s, to gain flexibility and reduce financial risks, most WCA countries reformed their pricing systems with a two-tier payment linked to world prices: a base price negotiated at the beginning of the cropping season and a price complement to be paid at the end of the season, if the cotton company makes a profit (Cameroon) or if the realized sales price is above the base price (Burkina Faso and Mali). Initial

51

cropping season prices, administratively set until the end of the 1990s, are now typically agreed on before the planting season through direct negotiation between cotton companies and farmers, or, in some countries (Burkina Faso and Mali) through a commonly agreed on pricing formula within an inter-professional committee. The parameters of the pricing formulas have been progressively improved: references to production cost (in Mali) have been abandoned, and replaced, in both Burkina Faso and Mali, by a fixed percentage (60 percent) of the free on board (FOB) price paid to farmers, thus linking producer prices to world market prices. Even when such mechanisms are well designed and applied, the sectors still face the uncertainty of what the actual world price will be nearly one year from the time the initial producer price has to be announced. This uncertainty creates significant financial risks for the cotton companies that have to buy at this preset price. If actual market prices are lower than this level, cotton companies will incur trading losses, eventually leading to the need for bailouts or financial restructuring.

In all countries, the pricing mechanism had been linked to a stabilization fund designed to limit fluctuations in producer prices and prevent prices from falling below a "minimum" level—at least until the 2004 crisis, when the world cotton price fell by 30 percent. Only one of these funds (in Cameroon) survived the crisis, but even it was exhausted in subsequent years. The collapse of the stabilization funds—caused by farmer prices well above what was feasible given world market prices—exacerbated the financial crises in the WCA cotton sectors, and required that the governments provide heavy financial support in Benin, Burkina Faso, and Mali.

**Table 5.1 Summary of Pricing Mechanisms in WCA Countries, 2006**

| Pricing element | Benin | Burkina Faso | Cameroon | Mali |
|---|---|---|---|---|
| Administered price? | Mixed | Yes | Yes | Yes |
| Panterritorial, panseasonal? | Yes | Yes | Yes | Yes |
| How set? | Government has arbitrary role | Negotiated within inter-professional committee | Negotiated within inter-professional committee | Negotiated within inter-professional committee |
| Announced before planting? | Yes | Yes | Yes | Yes |
| Adjusted before harvest? | Yes | No | Yes | No |
| Secondary payment after marketing? | Yes | Yes | Yes | Yes |
| Linked to Cotlook A Index?[a] | Yes in principle, not so clearly in practice | Yes | Yes in principle, not so clearly in practice | Yes |
| Stabilization fund? | Yes, but exhausted | Yes, but exhausted | Yes, but exhausted | Yes, but exhausted |
| Sector-wide deficits? | Yes | Yes (estimated €110 million, 2004/05–2005/06) | None through 2007 | Yes (estimated US$91 million for 2005 alone ) |

*Source:* Authors.

a. The Cotlook A Index is an index published by Cotlook Limited, an independent Merseyside-based company engaged in publishing cotton news for more than 75 years. The A Index is widely regarded as the reliable barometer of world cotton values, and is referred to by the major international cotton organizations, such as the ICAC.

Table 5.1 shows the essential elements of pricing mechanisms in the WCA study countries. The traditional cotton pricing mechanisms used in WCA have numerous economic and financial implications. First, panterritorial prices transfer resources from growers with low transport costs to growers more distant from gins or in less accessible areas. Second, when prices are announced before planting, price risk in the short run is borne by the cotton companies that will have problems operating profitably if market prices during the season fall below the level of the announced price. Longer term, taxpayers and donors also carry risk because the companies may need to be financially supported by the governments. Following the cotton price decline of the late 1980s, along with an overvalued CFA franc, several cotton companies had to be bailed out; the financial difficulties of most cotton companies in the region since the end of the 1990s are similar to the problems experienced at that time.

## Pricing Mechanisms in ESA

Before the reforms of the early 1990s, cotton pricing mechanisms in ESA closely resembled those of WCA in that a cotton parastatal (or cooperative unions in Tanzania and Uganda) was the sole buyer of cotton at a preannounced, panseasonal, and panterritorial price. Following reform, pricing mechanisms in the region became more market linked, flexible, and diverse, in line with the diverse sectoral types that have emerged (table 5.2). No country in the region operates a stabilization fund, nor have any generated sector-wide deficits that the government had to cover.

Mozambique operates the only local monopoly system in the region, and is the only country that maintains a fully administered, panseasonal, and panterritorial price. The government's role in price setting is strong in Mozambique, in part because of the very weak state of farmer organizations in the country. In Zambia and Zimbabwe's concentrated sectors, preplanting prices have been maintained, but this practice reflects business decisions by the dominant firms as they exercise price leadership in the sector; the government has no say in pricing in either country.[44] Prices paid to farmers throughout the region are much more strongly linked to medium-term, and even short-term, movements in the Cotlook A Index than they are in WCA. The way in which this happens varies greatly, however. For example, prices in Zambia largely adjust only year to year, as a result of price leadership by Dunavant, while in Tanzania and more recently in Zimbabwe (because of escalating inflation) they fluctuate throughout the marketing season. Even in Uganda's hybrid system, which attempts to eliminate competition among firms, prices vary over the course of the marketing season.

## Comparing Pricing Performance at the Farmer Level

Table 5.3 shows the share of the FOT lint price received by farmers in each of the study countries over the period 1990–2005. Producer prices for seed cotton are adjusted to lint equivalent using the average ginning outturn ratio, and any input costs borne by the companies are added to this result, to show the net value received by farmers. The Cotlook A Index is then adjusted to FOT based on transport and port cost data. Estimates of average quality premiums for each country (see chapter 7) are then added to derive the value received by the ginner at the ginnery door. The ratio of these two values—that paid to farmers by the ginners and that received by ginners at the factory door—shows the share of FOT paid to farmers. The FOT lint price is used instead of FOB because FOT is the final product price most within the companies'

**Table 5.2 Summary of Pricing Mechanisms in ESA Countries, 2006**

| Pricing element | Mozambique | Tanzania | Uganda | Zambia | Zimbabwe |
|---|---|---|---|---|---|
| Current structure | Local monopoly | Competitive | Hybrid | Concentrated | Concentrated |
| Administered price? | Yes | No | Only preplanting price | No | No |
| Panterritorial? | Yes | No | Only preplanting price | Yes for individual companies, but prices vary across companies | No |
| Panseasonal? | Yes | No | No | Yes | No |
| How set? | Negotiated between government and ginners; very little direct role of farmers | Competitive market price, no price leadership | Government (CDO) sets preplanting price in collaboration with ginners' association | Dunavant acts as price leader | Price leadership by Cottco and Cargill; newer companies generally pay more |
| Announced preplanting? | No | No | Yes | Yes, only by Dunavant | Yes (Cottco and Cargill only) |
| Adjusted before harvest? | n.a. | n.a. | Not formally, but actual prices paid do fluctuate over marketing season | Yes | Continually adjusted over season because of hyperinflation |
| Secondary payment after marketing? | No | No | No | No | Yes (Cottco and Cargill only) |
| Linked to Cotlook A Index?[a] | Yes in principle, not so clearly in practice | Yes | Yes | Yes | Yes |
| Stabilization fund? | No | No | No | No | No |
| Sector-wide deficits? | No | No | No | No | No |

*Source:* Authors.

*Note:* n.a. = Not applicable. CDO = Cotton development organization.

a. The Cotlook A Index is an index published by Cotlook Limited, an independent Merseyside-based company engaged in publishing cotton news for more than 75 years. The A Index is widely regarded as the reliable barometer of world cotton values, and is referred to by the major international cotton organizations, such as the ICAC.

control. Transport costs from FOT to FOB tend to be higher in landlocked countries (Burkina Faso, Mali, Uganda, Zambia, and Zimbabwe) than in coastal countries (Cameroon, Mozambique, and Tanzania). Thus, for example, costs from FOT to FOB are estimated to be 50 percent higher in Zimbabwe (US$0.157 per kg of lint) than in Tanzania (US$0.105 per kg of lint) entirely because of geography. (Because of different transport costs, seed cotton prices in US dollar terms may be lower in Uganda than in Tanzania, but the share of FOT paid to farmers may be larger.)

**Table 5.3 Summary of Producer Shares of FOT Lint Price**
*(percent)*

| | 1990–94 | 1995–99 | 2000–05 | 1995–2005 (post-reform in ESA) | Entire period (1990–2005) |
|---|---|---|---|---|---|
| Benin | 58 | 62 | 71 | 67 | 64 |
| Burkina Faso | 55 | 57 | 73 | 66 | 62 |
| Cameroon | 61 | 61 | 73 | 68 | 66 |
| Mali | 56 | 52 | 76 | 65 | 62 |
| Mozambique | 27 | 52 | 48 | 50 | 43 |
| Tanzania | 49 | 65 | 70 | 68 | 62 |
| Uganda | — | 72 | 68 | 70 | 70 |
| Zambia | — | 63 | 55 | 59 | 58 |
| Zimbabwe | 63 | 69 | 49 | 58 | 59 |

*Source:* Authors.
*Note:* — = Not available.

Caution is required in interpreting the data in table 5.3 for a number of reasons. First, the cotton companies in WCA, with the exception of Cameroon, have accumulated large deficits since the beginning of the 2000s, so that prices received by cotton growers include taxpayer (or donor) resources not captured by the FOT shares. Second, even FOT figures do not account for different rates of taxation across countries (for example, Tanzania taxes its cotton sector quite heavily while Uganda does not). Third, there are various exchange rate issues in six of the nine countries examined, implying that several of the cotton sectors have been taxed (the issues include, for example, a likely overvaluation of the CFA franc in WCA and the effects of local currency appreciation in Zambia). Finally, the ratio should be interpreted as an indicator of how well ginners are paying farmers compared with what they "should" be able to pay; it should not be seen as indicative of how well farmers are remunerated in an absolute sense. For example, average prices actually received by farmers have been very similar in Zambia and Tanzania; however, Tanzania's performance relative to FOT is much better than Zambia's because ginners in Zambia have generated a very high quality premium not enjoyed in Tanzania, but they pass little of this premium on to farmers.

## Conclusions

Focusing first on ESA, four patterns stand out. First, Mozambique (the region's only monopoly sector) paid extraordinarily low prices before 1995, in part as a result of additional costs that the companies had to bear: maintenance of private militias during the war, and substantial costs to keep roads open. As these costs disappeared during the final two periods, prices improved, but their FOT share remained the lowest in the region.

Second, and perhaps a surprise, FOT shares in Zambia and Zimbabwe (concentrated sectors) were relatively high in the five years following reform, even matching those in Tanzania and Uganda, which had more competitive sectors. Shares in both Zambia and Zimbabwe, however, dropped sharply during 2000 to 2005, clearly underperforming Tanzania and Uganda. We observe that the newly privatized sectors in both Zimbabwe and Zambia were making particular efforts to attract additional smallholders to cotton during 1995–99, while the fall in price shares during 2000–05 can be attributed to the fact that the sectors did not pass on to farmers the benefits of higher quality premiums on world markets (Zambia) or a major real exchange rate devaluation (Zimbabwe). The 1995–99 experience shows that farmers can receive reasonable prices under concentrated systems, while the 2000–05 experience shows that, in the absence of appropriate regulation, farmers are vulnerable to changes in the objectives or conduct of the dominant firms.

Third, considering the entire 10 years since 1995 (the post-reform era in ESA), Tanzania and Uganda clearly paid a higher share of FOT to farmers than any other country in the region. Both sectors have competitive structures. However, while competition remains unregulated in Tanzania, it has become highly regulated in Uganda since 2003. In Uganda, the continuing attractive prices are likely due to ginners' need to increase capacity utilization (very low at about 20 percent), their knowledge that farmers in Uganda move in and out of cotton based largely on relative prices (a dynamic seen much less in WCA), and therefore the need to pay attractive prices if the ginners are to attract growers. In Tanzania, over the past few years ginners have become more sophisticated regarding knowledge of global market prices and trends, ability to negotiate with buyers, understanding of price exposure, and the use of market-based approaches to mitigate that risk.

Finally, FOT price shares in WCA rose sharply through the 2000–05 period, reflecting the greater role of farmer organizations (supported by political pressures) in the price-setting process during that time, and the reluctance of stakeholders to reduce producer prices. In fact, over this period, price shares in every WCA country exceeded those in every ESA country. Clearly, however, these prices are not sustainable, as evidenced by the huge sectoral deficits generated in every country except Cameroon.[45]

Assessing the impact of price stability on rural income and poverty is clearly beyond the scope of this book. However, it is worth noting that more flexible prices are widely regarded in WCA as highly detrimental to rural development, poverty reduction, and development of the cotton sector.

# Chapter 6: Input Credit and Extension
## *David Tschirley*

Concern about input credit provision has long been at the center of debates regarding cotton sector reform in Sub-Saharan Africa (SSA). This concern is understandable in light of the widespread failure of input and, especially, credit markets in SSA.[46] The ability to supply large numbers of farmers with input credit and extension, and to recover that credit, was the driving force in the spectacular expansion of cotton production in West and Central Africa (WCA) from 1955 to 1995, made all the more impressive by the region's poor agro-ecological conditions; such provision has also been a necessary condition for reestablishing cotton in Mozambique after its civil war, and for the crop's rapid expansion in Zambia since 1994. The typology in chapter 4 was heavily informed by this issue, suggesting that more-concentrated sectors would be best able to ensure provision and repayment of input credit and some level of extension advice, while both of these would likely be undermined by side selling in more competitive sectors. It was further suggested that monopolies would perform well on input provision and repayment, while the adequacy and efficiency of the input and extension package might deteriorate over time by not adapting to changing conditions. The country case studies confirm these general hypotheses while providing detail and nuance that are relevant to sectoral policy discussions. This chapter briefly discusses approaches to and performance of input credit and extension across the eight countries,[47] organized around the typology of chapter 4, before closing with general lessons learned.

## Mali and Cameroon: Government Monopolies Show Similar Strengths and Weaknesses

Mali and Cameroon, along with Burkina Faso, share similar approaches to input credit and extension, inspired by the "West African model" of cotton promotion. The model features exhaustive coverage of farmers in agro-ecologically suitable areas with a standard in-kind credit package that includes about 50 kg of urea and 100 to 150 kg of compound fertilizer per hectare (ha), about six insecticide sprays per season, herbicides for some farmers, treated and reasonably well-maintained seed, and widespread adoption of animal traction. For many years, this input package was accompanied by a network of extension agents in charge of advising on agricultural best practices and technical itineraries for improved cotton cultivation. Most cotton sectors in WCA also now include an elaborate structure of village cotton farmer organizations configured into regional and national "apex" organizations. These apex organizations, along with ginning companies, form—or are in the process of forming— "inter-professional" bodies with responsibility to make joint decisions about input supply and pricing, among other factors. Notably, the government is not part of these inter-professional bodies, except through its ownership stake in the cotton company.

Objective indicators of input system performance are comparable across Mali and Cameroon:

- Input packages are nearly identical in composition and cost, and coverage of farmers is exhaustive in the main zones of each country.

- Average yields are about 200 kg (of seed cotton) per ha higher in Cameroon, but this may be due to agro-ecological factors more than to input quality (yields in the extreme north of Cameroon are nearly identical to those in Mali).

- Yields are trending down at about the same rate in each country, related at least in part to higher cost and thus lower use of fertilizers.

Two key differences in input and extension systems in the two countries are worth noting. First, Cameroon's producer organizations are substantially stronger and play an active role, in collaboration with SODECOTON, in input procurement, pricing, and distribution; credit recovery; and provision of extension services. Associations of larger, higher-yielding farmers in Cameroon employ and pay extension staff. OPCC, the national apex organization, employs 76 trainers to strengthen village-level associations. In Mali, the national apex organization does not yet exist, and it will likely take several years for the system to gain the financial and operational strength already seen in Cameroon.

A second key difference is that management of SODECOTON may be more attuned to opportunities to improve performance and reduce costs in its operations. One example is that it imports generic bulk insecticides that farmers mix in the field. SODECOTON claims that these products are cheaper, and that it is able to use them because of the relatively dense network of field agents, who are able to disseminate pesticide preparation techniques to farmers.

## Local Monopolies: Vastly Differing Histories Complicate Comparative Assessment in Mozambique and Burkina Faso

Burkina Faso moved to a local monopoly system with three firms in 2004, in the midst of a huge boom in cotton production made possible by many years of investment in research, input credit, and extension, as well as by relatively high prices paid to farmers since 2000.[48] In sharp contrast, Mozambique created its local monopoly system in the late 1980s, as civil war still raged and after national production had fallen below 10,000 tons of seed cotton. Even before the civil war and the economy's collapse, cotton production in Mozambique used far fewer external inputs than did the sector in Burkina Faso. One area in which the countries show similar performance is in the share of all farmers growing cotton in the cotton zones: 85 percent across Burkina Faso's whole cotton zone, and as high as 80 percent in Mozambique's cotton belt.

In reforming its cotton sector, Mozambique returned to the concession (or local monopoly) model prevalent during the colonial era. Key themes during the post-reform era have been the absence of any systematic approach to evaluating and re-awarding concession areas, extremely weak farmer organizations unable to negotiate with ginners or provide services themselves, widely divergent performance between early investors and new entrants (most of the latter affiliated with international cotton trading firms), recurrent credit default crises, and the government's openness to new investment, albeit always within the concession model. Until recently, the country clearly lagged behind its neighbors in productivity, though new entrants since the early 2000s have begun to change this in some areas of the country.

Key lessons from Mozambique's experience are, first, that a local monopoly system does not eliminate the possibility of credit default crises. If investment opportunities and regulatory capacity in a country are limited, the cotton sector is likely to attract new entrants, leading to side selling and increased credit default. Eventual decline in seasonal input credit and extension services is usually the result; in fact, both services are weakest, with extension almost nonexistent, in the areas most affected by credit default. Second, policy makers in local monopoly systems must select investors carefully. All companies in Mozambique face the same, very weak, government regulatory capacity. Yet some companies have chosen to invest in improved input supply and some extension, while others have operated for many years much like the new entrants in Zimbabwe, providing minimal quantities of poor quality input and little or no extension advice. Significant in Mozambique, the companies making the more aggressive investments are all affiliates of multinational cotton trading firms: Dunavant, Plexus, and CNA (affiliated with DAGRIS). These firms have all chosen to invest outside the traditional cotton growing zone of Nampula.

In Burkina Faso, the division of SOFITEX into three companies in 2004 changed very little with regard to input credit and extension; the West Africa model discussed above continues to be applied, though it seems likely that the severe financial difficulties of 2006 and 2007 have prevented companies from making much progress on their stated desires to modify input and extension packages. A key point to keep in mind as Burkina Faso moves down its reform path is that, despite the very developed structure and strong coverage of farmer groups within cotton areas, operational capacities remain weak. While UNPCB and its regional unions do receive and distribute input to members and organize the cotton market, neither is in a position to take over the importation and distribution of input to villages. Until this can happen, seasonal input credit from cotton companies will be critical to the sector's success.

## Competitive Sectors: Tanzania and Uganda Struggle and Take Very Different Paths to Ensure Input Supply, Extension, and Quality

The prereform, cooperative-based cotton systems in Tanzania and Uganda[49] transformed quickly after reform into highly competitive markets with 20 to 30 independent buyers vying for farmer production. Price competition was intense and farm prices improved, but each country witnessed the collapse of its input supply and extension system. As a result, the two countries in East and Southern Africa (ESA) that most closely approached the competitive ideal in market structure saw the most direct and persistent government involvement in efforts to ensure input provision to farmers.

Initial efforts in both countries involved sector-wide coordination of input provision—Tanzania's Cotton Development Fund (CDF) created in 1999, and Uganda's similar collaborative approach between ginners and the country's public Cotton Development Organization—so that ginners could concentrate on competition in the output market. Each approach achieved some success but failed after two years because of management and design problems.

Since these initial failed attempts, the two countries have moved in dramatically different directions. Starting about 2002, first informally and then through formal agreement, Uganda eliminated competition in the output market to facilitate input supply and extension by ginners (see Baffes 2008). Meanwhile, Tanzania maintained a competitive output market and implemented innovations in its approach to providing some minimal level of input to farmers (Poulton and Maro 2007). Uganda's zonal quota system featured collaborative production planning among two to three ginners in each of 11 zones, prohibited movement of seed cotton across zones, and facilitated sale of input at 50 percent of cost, with the subsidy implicitly collected in the price paid to farmers. Extension was a heavy focus in the system, with 7,000 demonstration plots and training days financed two-thirds by ginners and one-third by the US Agency for International Development. The consensus among ginners and observers was that input supply and extension would be drastically reduced if the zone system, or some variant of it, was not in place.

Despite these major efforts at input supply and extension, production in Uganda did not consistently rise above 20,000 to 25,000 tons of lint. In hindsight, it appears clear that a short-lived production boom in 2004 and 2005 was primarily due to high prices in the two preceding years (Baffes, background paper on Uganda, 2007). As a result of the inability to boost production, stakeholders in 2007 decided to abandon the zonal quota system, though it is not clear at this point what has replaced it.

Under Tanzania's passbook system introduced in 2003, farmers selling cotton receive a stamp in their passbooks that entitles them to seed or chemicals[50] the next year proportional to the amount of cotton they sold. For most farmers, the entitlement amounts to one or two chemical sprays and some seed (not enough for a full planting) the following year. The system is funded by a levy paid by ginners to CDF, which funds the importation of insecticides by private companies. Field interviews by Poulton and Maro (2007) suggest that the system has been "one contributory factor toward the major resurgence in cotton production in 2004 and 2005," but the authors conclude that "the system can make only a limited contribution to the intensification of cotton production in Tanzania," because it can finance only limited insecticide sprays and no fertilizer applications, and does nothing to provide extension assistance.

Geographical factors may provide some explanation for why these two countries, with very similar prereform histories and nearly identical structures after reform, chose (for a time) such different approaches to solving the input credit and extension problem. As long as the north of Uganda was closed, the scope for expanding cotton production under a low input approach was far lower than in Tanzania. Ginners in Uganda were thus forced to increase production in relatively small areas already under production. The fact that the sector abandoned the quota system at the same time that the north began to open up lends some support to this argument.

## Concentrated, Market-Based Sectors: Zimbabwe and Zambia Perform Well on Input Credit and Extension, but Face Instability

The two countries in ESA with single-channel marketing systems before reform maintained relatively concentrated sectors for several years after reform. Each has performed much better on input provision and extension than have Tanzania and Uganda. However, each has faced substantial structural instability that has affected both services.

Zimbabwe transitioned during the 1980s from a sector dominated by white commercial farmers to one with almost no such farmers, while building systems for effective input credit supply and extension assistance to a substantial minority of the new smallholder farmers. Cottco, the private company that emerged out of the government-owned Cotton Marketing Board with a market share of around 70 percent, continued this effective performance into at least the early 2000s and enjoyed credit repayment of 95 percent or higher in most years. Cargill, its main competitor, did not develop a credit system until 2002/03, but invested in extension support that encouraged loyalty from beneficiary farmers.

Between 2001 and 2006, the number of seed cotton buyers in Zimbabwe rose from 5 to 17, spurred by a fall in the real prices paid to farmers by the major players.[51] Credit default increased, and Cottco dramatically reduced input credit in 2004/05. Though the company has since expanded its system again, credit default remains a major problem. Draft regulations to deal with the situation were developed in 2004, but never enacted. During the 2006/07 season, the sector introduced a requirement that cotton companies must provide some input to producers to receive an export permit in future years.

An unusual result of Zimbabwe's move to a less concentrated system is that a substantially *larger* share of farmers received some form of input credit in 2006 than in the early 2000s. Whereas about 40 percent of growers received credit from Cottco or another company in 2002, nearly 95 percent received some type of support in 2006. However, regulation was a key driver of this result, and newer companies tend to provide seed of uncertain quality, little or no insecticide, and no extension advice. As noted above, the entry of these new companies was also accompanied by large increases in credit default among farmers. In an echo of patterns seen in Zambia and especially in Mozambique, widespread provision of very inadequate input packages (and no extension advice) has often been used as pretext to buy indiscriminately during the harvest. To the extent that this is happening in Zimbabwe, the apparent increase in credit provision may undermine such input credit and extension provision in the longer term.

Zambia's cotton sector built relatively effective input credit and extension systems in the years following reform in 1994, consistently providing farmers with high quality treated seed, four to six annual insecticide treatments, and (for the 20 percent or 30 percent of farmers that the main companies consider their best and most reliable) foliar feed fertilizers on 100 percent credit terms. Both major firms also stressed fundamentally sound agronomic practices with farmers, and Dunavant has attracted outside funding to support extension. Typical credit repayment was above 95 percent for Clark, and above 85 percent for Dunavant. As a result, the sector has seen slow but steady increases in the yields of established farmers and a near tripling of the total number of farmers growing cotton since 2000 (Tschirley, Zulu, and Shaffer 2004).

Despite the financial strength and high market shares of the two main companies, Zambia's cotton sector has experienced two severe credit default crises since reform. The crisis of 1998–2000 was overcome as Dunavant and Clark strengthened their input credit supply and recovery and (especially for Clark/Cargill[52]) extension systems, and demonstrated to most farmers the benefits of remaining loyal to the company. As a result, the credit default problem receded, and production boomed through the 2006 harvest season. The second credit default crisis occurred in 2006 and 2007, again spurred by the entry of new companies.[53] Unlike in 1998–2000, it appears likely that at least some of these new companies will be able to remain important players in the sector.

As in Zimbabwe, the sector is struggling to find a regulatory approach to deal with these stresses. Dunavant and Cargill, along with two of the emerging companies and farmers (represented by the Cotton Association of Zambia), are pushing for submission by the government to Parliament

## Box 6.1 The Instability of Concentrated Systems

***Concentrated sectors offer great advantages in SSA, but have tended to be unstable.*** The two concentrated sectors in this study—Zimbabwe and Zambia—have performed well on input credit and lint quality, but unevenly over time on seed cotton pricing. This suggests that, over the medium term in most of SSA, a concentrated sector is likely to be a more attractive option than national monopolies or highly competitive systems, especially if an appropriate regulatory framework can be put in place to encourage consistently remunerative pricing. Both Zambia and Zimbabwe have had difficulty maintaining concentrated structures since 2000; this appears to have been related to the pricing of seed cotton. In Zimbabwe, the duopoly that emerged out of reform in 1994 has, since 2002, been increasingly challenged by new competitors; these now number at least 15 and had a combined market share of 25 percent in 2006. Zambia also emerged from reform with a duopoly, went through one period of intense competition from new entrants in the late 1990s, and since 2006 has been in the midst of another such episode.

A move from a concentrated system toward a competitive one involves a qualitative change in the nature of competition within the sector: from competition for *producers* at the start of the season (with those producers then receiving a range of preharvest services to assist their production) to price competition for *seed cotton* at harvest time (with incentives for providing preharvest support undermined by side selling). In both Zimbabwe and Zambia, increased competition has undermined input credit provision; in Zimbabwe, it also undermined the country's previous reputation for high quality lint (chapter 7). Worse, competition may undermine input credit or lint quality— with long-term consequences for the industry—before it has any positive effect on prices paid to farmers. For example, there is little evidence that increased competition in Zimbabwe has yet improved the seed cotton prices received by the majority of farmers. Finally, while movement from a concentrated to a more competitive system can happen quickly in the absence of appropriate regulation (see next section of this box), reversing that change may be more difficult (see text of this chapter). If, as this study suggests, highly competitive sectors are likely to perform poorly in most of SSA, then such structural change may impose long-term losses on these countries.

***Inadequate regulatory structures are a key reason for instability.*** A key insight from the typology in this book is that the appropriate roles of ginners, farmers, and the government differ markedly across sector types. In concentrated systems, governments need to provide an environment in which the small number of buyers can coordinate input credit provision and quality enhancement, while ensuring that farmers receive remunerative prices. This could suggest that limited barriers to entry be established in the form of conditional buying licenses, such that only companies with a demonstrated commitment to productivity and quality should be allowed in. Such barriers would avoid many of the damaging consequences of uncontrolled new entry into concentrated sectors, but could also further reduce the effectiveness of new entry in raising seed cotton prices. For this reason, it is also suggested that more formalized approaches to price setting may be needed to ensure remunerative prices to farmers. In the latter, strengthened farmer associations would be much preferred to direct government involvement in price setting. In all cases, a government needs to carry out its role collaboratively with private investors and farmers.

Though governments in ESA have a limited history of working in this way, positive signs are emerging in both Zambia and Zimbabwe. In the former, the government worked closely with ginners and farmers to propose revisions to the Cotton Act, which is now awaiting submission to Parliament. If enacted, the act would establish a reasonably balanced public-private approach to sector management. In Zimbabwe, actors are considering a code of good conduct that cotton companies must adhere to if they are to receive an export permit.

This analysis suggests that lower prices (as a share of free on truck lint value) paid by existing firms have been the major factor encouraging new entry in Zimbabwe and Zambia. Because economies of scale in cotton processing are relatively modest (especially where Indian roller ginning technology is used), new investors can enter the sector quickly, even if they have no capacity or background in promoting input use and farm-level productivity. In addition, government policy sometimes promotes new investment in ginning without understanding the negative consequences this can bring. This is an emerging issue in Zambia, where the government has welcomed substantial new investment. The type of regulatory structure suggested above would establish standards other than the theoretical competitive ideal for evaluating whether such investment will be healthy for the sector.

of the revised Cotton Act, which would create a cotton board with power to regulate the sector but not to participate as a buyer or seller. Also as proposed in Zimbabwe, ginners and buyers would have to abide by specified rules of conduct to be granted a license, and could be subject to fines and seizure of cotton if shown to be involved in the promotion of side selling.

A key point emerging from this review is that, in the weak institutional and regulatory environment typical of SSA, concentrated, market-based systems may be unstable, with a recurring tendency to move to a more competitive structure (box 6.1). Tipping points may exist, in which the entry of two or three additional companies can dramatically change the prospects of coordination for input supply and extension (and quality control; see chapter 7). As the number of players rises, extension and input credit are the first services to suffer, certainly in quality (in Zimbabwe) and also in the number of farmers served (in Zambia). A key question that emerges is whether these systems will be successful in their efforts to establish enforceable rules of the game that allow enough competition to ensure good pricing performance (see chapter 5) while safeguarding credit repayment.

## Conclusions

Table 6.1 summarizes the assessment of input credit and extension performance across sector types. Realized performance matches expected performance most closely for competitive sectors and national monopolies. Uganda, which defied expectations for competitively structured systems, did so only by eliminating competition among firms, at least until 2007. Concentrated, market-based systems and local monopolies show the greatest diversity in performance. The former can perform quite well on input credit and relatively well on extension, but that performance can suddenly decline as a result of the instability referred to above. The sample of countries includes only one (Mozambique) where local monopolies had existed for a sufficiently long time to allow reasonable assessment. Highly variable performance across concession companies in Mozambique suggests that policy makers must carefully select new investors with an eye toward productivity and quality; affiliates of multinational cotton trading firms have performed best in this regard in Mozambique.

A key question that cuts across local monopolies and concentrated systems concerns the factors that favor or hinder the emergence of an effective regulatory approach featuring active collaboration among ginners and farmers, and balanced involvement of the government. The hypothesis is that firms are motivated to collaborate by both fear of loss and desire for gain, but that fear of loss may be the strongest motivator.[54] If this is the case, concentrated market-based systems, which provide the prospect of loss through competition, may provide better incentives for effective regulatory approaches. This expectation is conditioned by two factors. First, a history of collaborative decision making matters. Second, strong farmer organizations can impose losses on ginners under local monopolies by boycotting or otherwise confronting behavior to which they object. In both situations, countries in WCA are in a better position to achieve effective collaborative regulation under local monopoly set-ups than are countries in ESA.

# Table 6.1 Summary of Input Supply and Extension Systems

| Country | Current sector structure and governance | Mechanisms for extension and input credit supply | Percentage of cotton input sourced independently by farmers | Receiving some cotton input credit | Receiving some extension advice; quality of advice | Indicators — Using inorganic fertilizers | Adequacy and quality of package received on credit | Cost | Credit repayment rates |
|---|---|---|---|---|---|---|---|---|---|
| Mali | National monopoly | Extension and in-kind credit by CMDT to farmer through cooperatives; little operational involvement by farmers | Negligible | ≈ 100% cotton farmers, >90% all farmers in cotton zones | Provision comparable to input credit; quality declining over time | ≈ 100% | Treated seed, urea, compound, pesticides, some herbicides; questions about seed quality and appropriateness of fertilizer received | Standard package US$119/ha ("at cost") 35%–45% of mean production value | 95%; fell to 90% as early as 2001 |
| Cameroon | National monopoly | Extension and in-kind credit jointly managed & financed by SODECOTON & farmer apex; decreasing involvement of SODECOTON | Negligible, but reliance on SODECOTON is decreasing | ≈ 100% cotton farmers, >90% all farmers in cotton zones | Provision comparable to input credit; quality trend not clear | ≈ 100% | Treated seed, urea, compound, pesticides, some herbicides; little or no adjustment to differing agro-ecological conditions | Standard package US$123/ha ("at cost") 35%–45% of mean production value | 95%–99%; fell to 90% 2006 |
| Burkina Faso | Local monopoly | Extension and in-kind credit by 3 companies, with some limited operational involvement of farmer organizations | Negligible; intention to transfer task to farmers, but limited progress | ≈ 100% cotton farmers, 85% of all farmers across whole cotton zone | Provision comparable to input credit; quality trend not clear | ≈ 100% | Treated seed, urea, compound, pesticides, some herbicides; little or no adjustment to differing agro-ecological conditions | Standard package US$171/ha 45%–55% mean production value. Seed sold at 55% of cost. | 95% |
| Mozambique | Local monopoly | Extension and in-kind credit from ginning companies; highly variable quality; negligible involvement of farmer organizations | Negligible | ≈ 100% cotton farmers, >80% all farmers in key cotton zones | Negligible in Nampula (center of credit default problems); substantially higher outside Nampula | ≈ 0% | Highly variable across companies; mix of treated & untreated seed; some pesticides; little or no fertilizer | Highly variable. Typically US$10–US$30/ha, 10%–20% mean production value. Seed free. | Highly variable: 60%–90% |
| Zambia | Concentrated | Extension and in-kind credit from main ginning companies; no operational role to date for farmer organizations | Negligible | ≈ 100% cotton farmers, 30%–35% all farmers in cotton districts | Provision comparable to input credit; quality typically better from Cargill, though Dunavant attracts outside funding | 20%–30% (foliar only, though not just for micronutrients) | Treated seed, pesticides (5–6 sprays), some foliar fertilizer for 2 or 3 main companies; seed & limited pesticides from others | US$20–US$30/ha, 10%–20% mean production value; some evidence that sold above market rates | Typically 85%–98%. Falls below 70% during periodic crises |
| Zimbabwe | Concentrated (becoming competitive) | Extension and in-kind credit from ginners; highly variable quality; main companies (Cottco, Cargill) highly selective of best farmers, others get mostly poor farmers | Up to 60% early 2000s, now falling | 90%–95% cotton farmers (up from 40% early 2000s); 70%–80% of all farmers in main cotton zones | Primarily from Cottco and Cargill plus some from state before economic crisis; quality was high but more attention to loan recovery since new entry | 45%, covering nearly 90% of cotton area | Cottco: treated seed, fertilizers, chemicals; some newer companies: only seed and limited chemicals | Cottco: $237/ha, 43% mean production value Others: US$50–US$90/ha, 33%–39% mean production value | 90% Has fallen since early 2000s |
| Tanzania | Competitive | No extension or input credit; passbook system for input supply linked to "forced savings" | % outside passbook: seed 75%, chemicals 50%–75% | 0% | No private extension; public system drastically underfunded | 1%–2% | (Passbook, not credit) No treated seed; chemical quantities inadequate for full spraying regime (2–3 sprays for most farmers) | Seed US$1–US$2/ha; chemicals variable, depending on passbook entitlement & cash purchases | n.a. |
| Uganda | Hybrid (competitive) | Cash sale by ginners at 50% cost, implicit recovery of subsidy in price; extension covered jointly by ginners and donors | ≈ 20%; active secondary market in subsidized chemicals | (subsidized cash sale) ≈ 100% cotton farmers | Widespread extension and training organized around demonstration plots, jointly funded by ginners and donors | <10% | (implicit credit) Highly variable, as farmers free to purchase input they wish; all use treated seed; nearly all use some insecticides | US$6–US$8 for most farmers, 6%–18% of production value. | n.a. |

LESS COMPETITIVE ← ← ← ← ← ← ← ← ← MORE COMPETITIVE

*Source:* Author's compilation.

*Note:* n.a. = Not applicable. Uganda's structure is competitive but its conduct through 2007 was not, as a result of the regional quota system. Its cotton sector is therefore classified as a hybrid in this book.

# Chapter 7: Quality Control

## *Gérald Estur, Colin Poulton, and David Tschirley*

As discussed in chapter 2, the fiber characteristics of African cottons are typically superior to the Cotlook A Index. Because it is nearly all handpicked, African cotton is also cleaner and has fewer neps than cotton of most other origins. For these reasons, African cotton could command as much as a US$0.10/lb premium on international markets if the region could develop a reliable reputation for lint uncontaminated with foreign matter.[55] The typology suggests that concentrated systems will be best able to achieve this potential premium, whereas competitive systems are expected to show very limited ability to protect fiber quality, and thus will achieve the lowest quality premiums (table 7.1). The typology predicts medium performance for national and local monopolies, with the main point being that actual performance depends critically on management culture and, in local monopolies, regulatory effectiveness.

Underlying these predictions is an understanding that two conditions must be satisfied if a sector is to produce high quality lint. First, ginners must be able to control their supply chains to receive high quality seed cotton. Ginning can only protect or damage fiber quality; it cannot enhance it. Second, the farmers must have the incentives to achieve high quality lint. This chapter investigates the extent to which these conditions hold in the study countries.

The chapter first reviews common practices in each study country that affect cotton quality, then develops estimates of the key quality indicator for each country—the average realized premium achieved in international markets. It then examines performance across sector types and closes by comparing expected with realized performance.

## Impact of Quality on Export Prices

Cotton production has similar characteristics across Sub-Saharan African countries that impact quality. Upland cotton, grown on small-scale farms, is entirely rain fed. Production is labor intensive, using manual or ox-drawn implements and relatively few purchased inputs, and all seed cotton is harvested by hand. At national levels, African cotton is relatively homogeneous in fiber characteristics, as a result of similar growing conditions and the low number of varieties planted in most countries.[56] However, variability within bales is greater than in developed countries because the production of several farmers can be mixed in a single bale.

The price of African cotton on world markets is penalized by the way it is marketed and shipped. African shipments are less reliable, less homogeneous in quality and packaging, and have longer transit times than those of the major competitors from other world regions. Instrument testing is increasingly important in the global lint market, but most lint in Africa continues to be classified through visual and manual inspection, with instrument classification done only on a sample basis, if at all. International cotton merchants put great importance on the reliability of lint classification, independent of the actual quality of the lint; lint that is typically high quality but not reliably classified will not earn the premium that it otherwise would.

**Figure 7.1 Estimated Premium for Top Type of Lint during 2006/2007, by country**

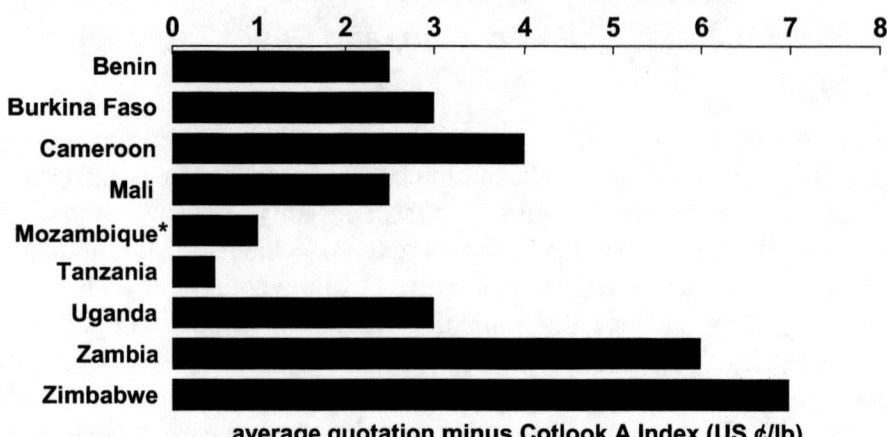

**average quotation minus Cotlook A Index (US ¢/lb)**

*Source:* Author's calculations based on quotations in Cotton Outlook and international traders' price estimates for Mozambique.

*Not quoted in Cotton Outlook.

Because African cottons show little variability in basic fiber parameters, price differentials between different origins primarily reflect their level of contamination (real or perceived). In 2006/07, the average premium of the quotation for the top type of each country in *Cotton Outlook* over the Cotlook A Index ranged from 1 to 7 US cents per pound (¢/lb), with Zimbabwe and Zambia receiving the highest premium and Tanzania and Mozambique the lowest (figure 7.1).

From the mid-1990s to the 2006/07 season, premiums for top types increased in Zambia (plus 5.0¢/lb), Cameroon (plus 4.5¢/lb), Mozambique (plus 2.0¢/lb), and Burkina Faso (plus 1.0¢/lb). Progress made in Zambia and Cameroon is due to successful reductions in contamination and stickiness, respectively. In Mozambique, newer concessionaire companies, such as Plexus and Dunavant, provided high quality cultures that were lacking in the older companies. In contrast, differentials for top types against the Cotlook A Index declined by 1.5¢/lb in Tanzania, 1.0¢/lb in Uganda and Zimbabwe, and 0.5¢/lb in Benin and Mali. The declines reflect increased competition between ginners in all three East and Southern Africa (ESA) countries, and lax seed cotton grading and contamination resulting from poor management in the two West and Central Africa (WCA) countries.

The premium paid for the top types in each country does not necessarily reflect the overall effectiveness of quality control in those systems, because it does not indicate the share of those types in total production. Calculating an average realized price—which would more accurately reflect the success of quality control in the sectors—requires data on the share of each type in total production and the premium received by each of these types. Such data are exceptionally difficult to obtain. Therefore, a theoretical average quotation by country[57] is calculated based on the following data:

- average premium for the quotation of the top type, as shown in figure 7.1

- usual world market price differences for grade compared with Middling, as follows: Good Middling, plus 1.5–2.5¢/lb; Strict Middling, plus 0.75–1.0¢/lb; Middling, 0; and Strict Low Middling, minus 0.5–2.0¢/lb

- usual world market price differences for staple length, relative to $1^3/_{32}$ of
  - $1^5/_{32}$ inches: plus 1.5–2.0¢/lb
  - $1^1/_8$ inches: plus 0.5–1.0¢/lb
  - $1^3/_{32}$ inches: 0
  - $1^1/_{16}$ inches: minus 1.75–4.0¢/lb

- actual 2005/06 classing data for WCA countries and most recent available data or estimates for ESA countries

To calculate the theoretical average export price of the crop, a deduction of one cent per pound is applied to the weighted average quotation to reflect the usual difference between the seller's offering price and the actual negotiated contract price. Based on these calculations, theoretical weighted average export price differentials compared with the Cotlook A Index range from minus US$0.02/lb to plus US$0.04/lb (figure 7.2).[58]

## Impact of Sector Organization on Quality

Zambia's concentrated sector stands out as the best performer in figure 7.2. It has also seen the greatest increase in premium since the mid-1990s. The reduction in lint contamination over this period was achieved through strict control over the quality of seed cotton admitted to the ginneries of the two dominant companies. Dunavant installed cleaning stations at its buying posts, at which all seed cotton is hand sorted before being sent to the ginnery. The company has also refused to accept seed cotton brought to buying posts in polypropylene bags. Moreover, it has introduced a third grade for seed cotton purchasing (A+) that rewards farmers who deliver contamination-free lint. Similarly, Clark/Cargill sought to instill in the farmers it works with the importance of delivering high quality seed cotton. As a result, farmers who disregard quality or who seek to adulterate their seed cotton have few outlets at which to sell it. Both Dunavant and Clark/Cargill are directly engaged in lint marketing to regional or international spinners, so are fully aware of, and capture, the benefits of high quality lint. This, combined with their control over their supply chains, lies behind the strong quality performance observed in the Zambian sector.

**Figure 7.2 Estimated Premium Weighted Average Basis - by country, US cents/lb**

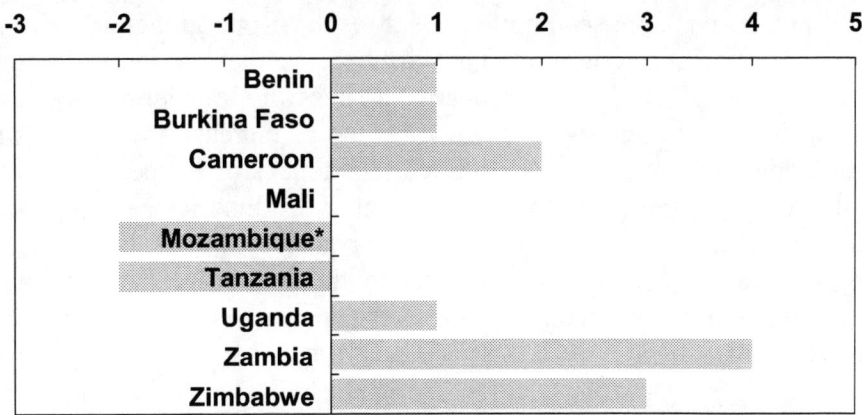

theoretical average price minus Cotlook A Index (US ¢/lb)

*Source:* Author's calculations based on 2005/06 season classing data by country, quotations in Cotton Outlook, and international traders' price estimates for Mozambique.
*Not quoted in Cotton Outlook.

National and local monopoly systems exhibited varying performance since the mid-1990s. Within WCA, Cameroon shows what a national monopoly can do when political interference is kept to a minimum and the company is left to run along fairly strict commercial lines. By contrast, Mali's zero premium is unimpressive when one recalls that its fiber characteristics—like those of nearly all fiber coming from Africa—are superior to the Cotlook A Index. Political interference has played some role in Mali's poor performance. Mozambique's poor performance reflects the legacy of a nearly unregulated local monopoly system with original concession companies uncommitted to productivity and quality; while quality is likely better among the newer affiliated ginners (Dunavant, Plexus, CNA/DAGRIS), it will take time for them to overcome the country's poor reputation.

In theory, monopoly systems should be able to exert the same control over their supply chains as concentrated sectors. If they take a hard line on the quality of seed cotton received, penalizing poor quality through price discounts or, in extreme cases, refusal to accept delivery at all, farmers have no alternative outlet at which to sell. However, the social and political context is also important. Unlike in concentrated sectors, where the notion of competition is extremely helpful in this regard, there is an expectation that the companies, as monopolies, will accept all seed cotton delivered to them. This is particularly strong within the national monopoly system in Mali, where responsibility for grading has been transferred to producers' associations. As a result, grading is very lax. There are large stated price differentials between grades, but these result in 99 percent of the crop being purchased as 1st grade regardless of the subsequent lint classifying results. Contamination is often not taken into account and little care is given to the cleanliness of cotton before it reaches the gin.

The effect of Tanzania's competitive system is clearly seen in the decline in its premium since liberalization and its resulting 2¢/lb average discount relative to the Cotlook A Index (figure 7.2). The negative impact of competition on cotton quality in Tanzania was first highlighted by Gibbon (1999). Unregulated competition undermines the ability of ginners to control their supply chains, while the limited vertical coordination between the large number of small, independent ginners and the country's lint exporters weakens the incentives that ginners face to produce high quality lint. For most of the postliberalization period (excluding the bumper years 2004 and 2005), the large number of ginners was associated with serious excess ginning capacity. Thus, ginners scrambled to buy available seed cotton. As a result, if one ginner seeks to impose strict grading requirements during seed cotton buying, farmers take their cotton to a competing buyer who is more lenient.[59] The laxity of grading means that unscrupulous farmers can also adulterate their seed cotton with sand, water, rocks, or other items to increase the weight of their bales. Farmers claim that they do this in response to buyers' practices of tampering with the weighing scales. Meanwhile, an exporter buying lint from a number of ginneries cannot be sure of getting a higher price for better quality, especially if selling forward. However, if exporters do obtain a good price, they do not necessarily pass that on to the ginner in question. Many ginners, therefore, place more emphasis on increasing turnover than on raising quality. Box 7.1 explores these issues in more detail.

While Uganda has a large number of ginners, as does Tanzania,[60] the Ugandan sector is heavily regulated, with a quota system for seed cotton purchase eliminating the direct competition for cotton that has been so detrimental to quality in Tanzania. Since the mid-1990s, Uganda's premium has nevertheless fallen, although it remains above Tanzania's. Some of this difference

**Box 7.1 Why is Quality Management So Hard within Competitive Sectors?**

It is estimated that the efforts made by the two main companies in the Zambian cotton sector to control contamination in seed cotton cost about 1¢/lb of lint produced, whereas the benefit has been an increase in the premium of around 5¢/lb of lint since the mid-1990s. In this case, quality pays. In Tanzania and Zimbabwe, one can argue that small, independent ginners may not be rewarded by exporters for delivering higher quality lint. However, in both countries' sectors, ginners that are part of vertically integrated companies would benefit from being able to sell higher quality lint. But experience in both countries indicates that quality-conscious cotton buyers are unable to unilaterally insist on grading and associated price differentials at primary marketing in highly competitive markets because they are undercut by competitors that put quantity before quality. The question, therefore, arises: why would higher prices achieved on the world lint market not allow these quality-conscious firms to pay higher prices for good quality seed cotton?

One answer is that competitors may achieve higher capacity utilization at their ginneries by being less selective in their seed cotton purchases, thus boosting the prices that they can pay. Interviews with the manager of the most quality-conscious cotton company in Tanzania indicated that (at least, before the bumper harvest in 2004) their insistence on quality control had caused them to forgo seed cotton volume. This then focuses attention on the size of the premium obtained by selling a higher quality product in relation to the cost savings from achieving higher capacity utilization. While cotton lint quality is important, the price differentials for lint of different qualities are not of the magnitude of those witnessed in, say, coffee.

Furthermore, there may be other reasons—generally but not always of a transitory nature—that distort competition between quality-conscious buyers and others. For example, widespread evasion of the high taxes and levies imposed on cotton buying and ginning in Tanzania is alleged. However, companies with an international brand reputation (which tend to be among the more quality conscious) perceive it as too risky to their wider brand reputation to be caught in flagrant tax or levy avoidance. They may, therefore, end up paying more in levies and taxes than some of their less quality-conscious competitors. Similarly, in Zimbabwe in recent years, Cottco, the main company, argued that it was unable to match the prices offered by new entrants, even though it had a long-established reputation for quality on international markets, because it was much more visible to government agencies that were seeking to stamp out "abuse" of the rampant parallel foreign exchange market that developed in the country after 2001. In a similar vein, Tollens and Gilbert (2003) argue that shortages of foreign exchange in Nigeria in 1986/87 caused traders to scramble for cocoa in the newly liberalized market, sacrificing quality in the process. The parallel market mark-up on the scarce foreign exchange that could be generated through cocoa export exceeded the quality premium obtained from higher quality cocoa. Tollens and Gilbert (2003) argue that the Nigeria cocoa case is a special case. However, the Zimbabwe and Tanzania cotton cases suggest that these "special cases" may actually be quite common.

is due to the larger share of roller-ginned cotton in Uganda. However, another factor may be the much smaller size of the sector, which allows Uganda's Cotton Development Organization to monitor grading and other quality practices in a way that the Tanzania Cotton Board cannot hope to do. In Tanzania, there are in excess of 5,000 registered buying posts spread over a huge area of the country.

Finally, Zimbabwe appears to perform well, according to figures 7.1 and 7.2. However, this is a legacy of the outstanding performance delivered first by the national monopoly Cotton Marketing Board, then by the concentrated system through 2001. Following the rapid entry of new firms since 2001, system-wide quality control has suffered, even though Cottco and Cargill, firms with an established reputation for quality consciousness, still account for 70–80 percent of the market. In 2002/03 one of the new companies in Zimbabwe (from Tanzania) was the first to offer flat rate prices for all its seed cotton purchases (irrespective of grade) and even the established companies felt obliged to follow suit. During 2003/04 and subsequent seasons, the majority of primary marketing transactions were completed either without grading or with grading being merely a formality from the farmer's perspective because of the flat-rate pricing. Almost immediately, the average quality of seed cotton delivered by farmers declined because they no longer felt the need to grade their cotton.[61]

During fieldwork in Zimbabwe, Cargill reported that, in the mid-1990s, around two-thirds of all seed cotton would have received an A or B grade. In turn, all grade A and B cotton would have fed through to lint of the top three grades. In 2006, according to buying slips, 35 percent of Cargill's seed cotton received an A or B grade, but more rigorous regarding at the company's ginneries reduced this figure to less than 1 percent. As a result, even with the use of lint cleaners, only 3 percent of lint achieved the top grade (compared with 20–25 percent in the late 1990s) and less than 50 percent could be sold as one of the top three grades. Similarly, at Cottco's Gokwe ginnery, a striking decline was recorded in the proportion of seed cotton supplies classed as grade A between 2003 (26 percent) and 2004 (5 percent). Because 2003 was the year in which flat-rate buying was introduced within the sector, this dramatic decline in quality shows that farmers adapted to the new incentive system in a single season. Another notable drop in quality (principally affecting the proportion of B grade cotton) occurred in 2006, a year of exceptional price competition between companies. Thus, the share of seed cotton achieving grades A or B fell from 65 percent in 2003 to 20 percent in 2006. Returning to figures 7.1 and 7.2, the country's premium would have been substantially higher in the early 2000s, before the intensified competition within the sector, and is likely to fall in the latter years of the 2000s if quality control measures are not improved.

## Conclusions

Overall, quality performance is remarkably consistent with the expectations generated from the typology: Zambia's concentrated system delivers the best performance, comparable to what Zimbabwe delivered before the entry of new competitors. Zimbabwe's premium is still high, but falling as a result of increased competition. Cameroon, with a national monopoly largely free of political meddling, also performs well, though not up to the standards of Zambia. The other national and local monopolies show highly variable performance, while Tanzania's competitive system is, along with Mozambique, the worst quality performer.

Table 7.1 summarizes available quality information for each country in this study. Growing fewer varieties in a country makes it easier to maintain homogeneity of quality, though proper controls (as in Zambia, which grows at least two varieties) and good classification can ensure good performance even when several varieties are grown. Having more seed cotton grades is generally good, but only if strictly linked to lint classification outcomes. Instrument testing is increasingly important in the global lint market but is rarely used in Africa. International cotton merchants put great importance on the reliability of lint classification in a country, independent of the actual quality of the lint. Lint that is typically high quality but not reliably classified will not earn the premium that it otherwise would. Longer staple length is always good, and 1⅛ inch is a typical benchmark. Contamination is crucial in pricing, and a reputation for high contamination is difficult to overcome. All of these factors contribute to the average premium a country is able to earn over the Cotlook A Index; these estimates are presented in the final column of table 7.1, and discussed in more detail in the next chapters.

**Table 7.1 Summary Information on Quality Control Mechanisms and Results in Study Countries**

| Country | Sector type | Number of varieties | Number of seed cotton grades | Strictness of seed cotton grading | Share of lint classed by instrument testing (percent) | Classification rating | Share of lint classed ≥ 1⅛ inches and above (percent) | Reputation for contamination | Overall reputation and trend | Estimated weighted average premium over Cotlook A Index (US¢/lb) |
|---|---|---|---|---|---|---|---|---|---|---|
| Benin | Hybrid | 1 | 2 | Lax | 5 | Average | 76 | Moderately contaminated | Good but irregular | +1 |
| Burkina Faso | Local monopoly | 3 | 3 | Lax | n.a. (sample basis) | Good | 80 | Moderately contaminated and improving | Good, improving | +1 |
| Cameroon | National monopoly | 2 | 2 | Strict | 0, but micronaire tests for each bale | Good | 65 | Among most affected by stickiness but improving sharply | Good, improving (entered fine cotton market segment) | +2 |
| Mali | National monopoly | 2 | 3 | Very lax | 6 | Average | 98 | Among the most contaminated | Average, improving | 0 |
| Mozambique | Local monopoly | 8 | 2 | Lax | 0 | Poor | ≈ 15–20 | Moderately contaminated | Poor, possibly improving | −2 |
| Tanzania | Competitive | 1 | 2 | Very lax | n.a. (sample basis) | Average | ≈ 30–40 | Among the most contaminated | Poor, fell since reform | −2 |
| Uganda | Hybrid (competitive structure) | 1 | 2 | Very lax | n.a. (sample basis) | Average | 93 | Among the most contaminated | High but much lower than in 1970s | +1 |
| Zambia | Concentrated | 3 | 3 | Strict | 70–80 | Very good | ≈ 70–80 | Very good | High and improving | +4 |
| Zimbabwe | Concentrated (becoming competitive) | 2 | 4 | Very lax | n.a. (sample basis) | Mixed | ≈ 70–80 | Moderately contaminated | Fell sharply since 2002 | +3 |

*Sources:* SONAPRA, SOFITEX, SODECOTON, CMDT, IAM, TCA, CDO, Dunavant, C. Poulton, international traders.

*Note:* n.a. = Not applicable.

71

# Chapter 8: Valorization of Seed Cotton By-Products

## Nicolas Gergely and Colin Poulton

As pointed out in chapter 2, by-products derived from cotton seed processing have growing markets and are an important complementary source of revenue for cotton sectors in Africa. According to standards worldwide, 20–25 percent of the total value of seed cotton can come from the by-products (cottonseed oil and cake). However, one stark finding in this study is that the markets for these products are not well developed in Africa and prices received by ginners are often low; thus, farmers are generally not getting the full returns from the processing of cotton seeds.

## Structure and Organization of Oil Sectors

The study countries show important differences in the organization and performance of their cottonseed oil sectors. These differences exhibit some parallels with the differences in organization across the cotton sectors. However, because the organization of the oil sector is not the major focus of this book, it will only be discussed briefly. Meanwhile, differences in the performance of oil sectors have important consequences for the profitability and competitiveness of each national cotton sector.

Table 8.1 shows considerable variation in the number of cottonseed oil processors across the study countries. There is no domestic industry in Mozambique (where demand for cake is extremely limited), a monopoly in Cameroon, one dominant price setter in both Burkina Faso and Mali, and an increasingly competitive market in some of the other countries. Vertical relationships cover a wide spectrum, from full integration with cotton ginning (as in Cameroon, most cottonseed oil processors in Tanzania, and three new companies in Zimbabwe), through various forms of vertical coordination (SN SITEC in Burkina Faso and larger companies in Zimbabwe), to market relationships.

Transaction cost economics (Williamson 1985) can provide insights into the varying vertical relationships. In an African context, large-scale processing equipment for high quality oil (deodorized, neutralized, cleaned, with gossypol removed) is a fairly specific asset. In Africa, either vertical integration or coordination could strengthen the incentives for investment in oil processing. Thus, in Zimbabwe three large-scale oil processors have maintained close relationships with the two largest cotton companies through regular interaction on the National Cotton Council (NCC).[62] By contrast, outside of West and Central Africa (WCA) and Zimbabwe, processing operations are much smaller and often produce lower quality oil. Many of the new processing plants being installed in Burkina Faso, Mali, Tanzania, and Zimbabwe use low-cost Indian equipment. For such plants, relying on spot market purchases of seed presents only a modest risk.

However, the main factor explaining observed variations in sector structure is not the techno-economic attributes of different types of processing equipment, but policy choice. Until the 1990s, in WCA countries vertical integration of large-scale oil processing within the parastatal

# Table 8.1 Cotton Seed Production and Processing in Study Countries

| Indicator | Benin | Burkina Faso | Cameroon | Mali | Mozambique | Tanzania | Uganda | Zambia | Zimbabwe |
|---|---|---|---|---|---|---|---|---|---|
| Average national seed cotton production (tons) 2001–06 | 339,500 | 557,833 | 242,966 | 488,281 | 72,178 | 235,000 | 78,410 | 160,000 | 246,350 |
| Cottonseed oil production as percentage of national oil consumption[a] | 53 | 57 | 18 | 50 | 6 (potential) | 8 | 4 | 20 | 27 |
| Number of cottonseed oil processors (2006) | 2 | 11 | 1 | 2 | 0 | 13 or more | 4 | 3 | 6 or more |
| Retail price of oil (US$/liter) 2006–07 | — | 0.95[b] | 1.28 | — | — | 0.83[b] | — | 1.29 | 1.55 |
| Seed price (US$/ton) 2006 | 63 | 44 | 59[c] | 50 | 55 | 27–117[d] | 86 | 71 | 95 |
| Landlocked | no | yes | no | yes | no | no | yes | yes | yes |

*Source:* Authors.

*Note:* — = Not available.

a. Estimated from the quantity of seed available for crushing (2001–06 average), after subtraction of seed retained for redistribution to farmers, using an oil outturn of 18 percent and an average annual oil consumption of 7 kg per person.

b. Some of these prices have surged in 2007/08, such as the retail price of cotton oil in Burkina Faso which went up to US$2.00 per liter.

c. Tanzania figure is wholesale price, so estimated retail price = US$1.00 to US$1.08 per liter.

d. Accounting price recorded within the integrated company.

e. US$27 per ton is lowest price reached in 2005 (bumper harvest), while US$117 per ton is highest price reached during 2006 (drought year). Average figure is not available, but some respondents reported that a "normal" price would be about US$50.

cotton company was part of the development model for the national cotton sector. In subsequent years, an early part of the reform process was the privatization of the oil processing enterprises. These privatizations have rarely been open and transparent, however. In Mali, a private processing monopoly was created by the divestment of the state from HUICOMA. The new company has performed so far poorly and is now in dire financial straits, with smaller operators entering the market in competition. In Burkina Faso, the French company DAGRIS, a shareholder in the main cotton company SOFITEX, is also the majority shareholder in the main private oil processor. Cameroon, the one country in the sample where little significant reform of the cotton sector has taken place, is also the one country where ginning and oil processing have remained integrated. Oil processing in Cameroon is managed in an entrepreneurial manner and contributes to the overall stability of the cotton operation.

In East and Southern Africa (ESA), where the cotton market has been liberalized since 1994/95 (earlier in Mozambique), oil processing is entirely in private hands. Given the small proportion of national edible oil consumption that cottonseed oil supplies in these ESA countries (table 8.1), its processing has limited strategic importance. As a general rule, where multinational trading companies have invested in cotton production in these countries, they have shown little interest in cottonseed oil processing because volumes are too small for companies primarily interested in international markets. Oil processing has thus been left to domestic or Asian entrepreneurs.

In Tanzania, the first private oil processors entered the cotton sector soon after liberalization as a way to guarantee access to seed supplies. Several cotton ginners recently entered into oil processing, although it is not clear whether these ginners are seeking to stabilize the price they realize from their seed (see the recent price fluctuations reported in table 8.1) or are responding to the attractive profits obtained by the existing oil processors.

In Zimbabwe, investment in oil processing equipment by Indian- and Tanzanian-owned cotton companies has been observed during 2005–07. This is a response to a national shortage of edible oil, part of the ongoing economic crisis in the country. Until 2000, the three established oil processors supplied about 80 percent of national oil requirements using a blend of soybean oil and cottonseed oil in roughly equal proportions. With the onset of the fast-track land redistribution program, however, soy production contracted rapidly, while cotton production has recently been lower than it was during 1999–2001. The shortage of local raw materials has been compounded by a lack of foreign exchange for either imported materials or edible oil. Thus, attractive profits can be obtained by those who control the supply of scarce cotton seeds.

In general in Africa (except possibly South Africa) farmers do not own the cotton seed. To "own the cotton seed," farmers either need to own ginneries or to toll gin. The latter option is

### Box 8.1 Toll Ginning in Zimbabwe

In Zimbabwe, Cottrade offered a brokerage service for farmers during the period 2000–04 whereby they arranged toll ginning contracts and assistance with the sale of both lint and seed, for a 2 percent commission. With the exit of commercial farmers, Cottrade tried to work with groups of smaller producers. In the years when Cottco and Cargill failed to pass the benefits of exchange rate depreciation on to producers, organized producers could get much more money via Cottrade than through normal channels. However, few were sufficiently organized or could produce the necessary volumes without company credit. In 2004, the exchange rate stabilized and the gap between what farmers could achieve through Cottrade and normal channels narrowed significantly. This seems to be what persuaded Cottrade to cease brokerage operations, but lack of progress with farmer organization may have been a contributory factor. In 2006, Cottrade began operating more like a traditional company in Zimbabwe, purchasing seed cotton and processing it.

easier, but still requires significant volumes per consignment—mandating a degree of farmer organization that is lacking in most countries. Farmers also need to be able to finance their own production without cotton company credit, thus remaining free of obligations to these companies. However, in Zimbabwe a company called Cottrade did set up a toll ginner operation (see box 8.1). Even though Cottrade's experience lasted only a few years, it indicates that some of the obstacles mentioned above could be overcome and that new ways of doing business could emerge.

## Performance of Oil Sectors, Seed Pricing, and Returns to Farmers

For the purposes of this study, a key performance indicator for the oil sector is the price that ginners pay (or receive) for cotton seed. Given the competition from imported palm oil in all the study countries, the main determinants of the cotton seed price are expected to be the following:

- Whether the country is landlocked (a major determinant of the overland transport cost incurred by imported palm oil). Note that this partially compensates ginning companies in landlocked countries for the high free-on-truck minus free-on-board costs they incur when exporting their lint.

- The level of tariff protection (if any) offered to domestically produced edible oils. In Benin, Burkina Faso, and Mali, a common external tariff of 5 percent applies to edible oil imports. However, some smuggled oil imports avoid this tariff and also avoid the 20 percent value added tax that is applied to domestically produced oils. Thus, domestically produced oils are at a net disadvantage when compared with these smuggled oils.

- Whether oil processors are able to brand cottonseed oil, so as to raise its price above that of imported palm oil.

- The quality of oil produced, insofar as this reflects the degree of processing. Discussions in Tanzania suggest that it could cost twice as much to produce high quality oil (deodorized, neutralized, cleaned, and gossypol removed) as to produce a semi-refined oil.

- The strength of demand for cottonseed cake from the domestic livestock industry.

- The efficiency of the oil processing sector.

A recent survey by the International Cotton Advisory Council (ICAC) provides useful information with which to benchmark the valorization of cotton seeds with world market standards. According to the ICAC survey, globally a kilogram of seeds fetches on average 18 US cents/kg, which represents a good income for the grower. The data by region show that cotton seed has a higher value in non-WCA African countries and in Asia, where a kilogram is sold at 22 cents/kg and 20 cents/kg, respectively. A kilogram of cotton seed after ginning is sold at 13 cents in North America (average of Mexico and the United States) and 10 cents in South America. Cotton seed prices are the lowest in West African countries, where a kilogram of seed is sold at 7 cents (Chaudhry 2007).

Table 8.1 shows the cotton seed prices paid by oil processors in the nine study countries in 2006. The table shows that the price paid in Zimbabwe was more than twice the price paid in Burkina Faso (the lowest recorded within the group). This degree of variation cannot be justified by economic fundamentals. The following can be observed:

- Landlocked countries in East and Southern Africa (ESA; Uganda, Zambia, and Zimbabwe) record high prices for seed, as expected. In Zimbabwe, the observed price is higher than otherwise, most likely a reflection of the current exacerbated oil shortage.

- Coastal countries in both ESA and WCA record modest prices for seed, again as expected. A range of US$55–US$63 per ton is observed, with Benin the highest, followed by Cameroon. However, as noted in table 8.1, the price in Tanzania can go well above this price range when seed supplies are limited.

- It is more difficult to brand cottonseed oil as a superior product (commanding a price premium over imported palm oil) when it accounts for the majority of the total oil market than when it accounts for a much smaller share. Within the four WCA countries, the only one where cottonseed oil has been successfully branded and promoted is Cameroon. Likewise, in Tanzania some of the early private oil processors established brands that are well known within the cotton-growing regions around Lake Victoria. Cottonseed oil is preferred over palm oil for the frying of fish and doughnuts because it burns at a higher temperature.

- Demand for cottonseed cake from the domestic livestock industry is much stronger in the Sahelian countries (Burkina Faso and Mali), plus possibly Zimbabwe,[63] than in the other countries in the sample. In the Sahelian countries, cake is sold for around CFA franc 50/kg (US$0.1/kg) ex factory, which makes the value of cake about half that of oil (given a cake outturn of 80 percent per kg of seed processed).

- Oil produced by the established companies in the four WCA countries plus Zimbabwe is of a higher quality (with commensurately higher processing costs) than that produced in Tanzania or by the newer companies in WCA and Zimbabwe.

Taking these observations together, the clear outliers in cotton seed pricing are Burkina Faso and Mali. Both are landlocked and exhibit strong demand for cottonseed cake, but in 2006 seed prices were only US$44/ton and US$50/ton, respectively. The retail oil price in Burkina Faso is also low for a landlocked country, which may reflect the impact of oil smuggled in through Togo. However, in both Burkina Faso and Mali, the fundamental problem appears to be monopsony power. In Mali, where the privatization of the parastatal processor has already been commented upon, even US$50/ton was an improvement over the price paid in the previous two years. In Burkina Faso, the privatized former parastatal SN SITEC still accounts for around two-thirds of seed purchases and acts as a price leader. The 10 smaller new entrants should eventually make a difference to the seed price, but currently do not appear to account for enough of the market (or have enough working capital) to push the price up significantly.

Overall, a limited, but variable, degree of development of domestic oil and cake markets is observed. From the perspective of cotton ginners, the main problem is the monopsony power exercised by oil processors in Burkina Faso and Mali, as well as a lack of transparency and contestability on the transfer price of cotton seeds established through long-standing arrangements between the cotton company and the oil processor—the latter often a subsidiary of the former—officially justified by the need to protect domestic industries. Nevertheless, oil consumers may gain at the expense of seed cotton producers in Burkina Faso.

Given the often tight margins obtainable from lint production, more research would be worthwhile on measures to improve the efficiency of the markets for oil and cake. Based on crude estimates of oil processing costs, there are reasons to believe that many of the observed seed prices could be raised by additional competition. On the consumer side, this does raise a quality issue: oils from newer, smaller processing units are typically less refined than oils from larger, established companies. However, as long as basic food safety requirements are met, it may well be that a significant proportion of poor consumers would willingly accept lower quality oil if its price was also lower. In this regard, there would appear to be parallels with the liberalization of maize markets in ESA in the 1980s, leading to rapid new entry of small-scale hammer mills in competition with established, large-scale roller mills (Jayne et al. 1995).

The other major challenge for the regulation of liberalized oil markets is the enforcement of tariffs and other taxes on imported oils. Cottonseed oil produced locally suffers in a number of African countries from unfair competition from massive imports of vegetable oils originating from Southeast Asia that are often imported fraudulently or without paying the full amount of legally due tariffs and taxes.

## Summary

This analysis reveals that there are very significant differences in the valorization of cotton seeds. Burkina Faso and Mali are landlocked, are deficit in edible oil, and have strong demand for cake from local livestock sectors, yet prices paid to cotton companies for seed are the lowest of the nine countries because of monopsony power in the oil sectors, whether public or private. Uganda's competitive oil sector makes a significant contribution to ginning company profitability as a result of the high prices paid for cotton seed. Small-scale investment in Tanzania has also led to more attractive prices for cotton seed. Similar investment is starting to take place in Mali, and should not be discouraged.

Globally, changes in institutional structure can be expected in many countries, resulting from the growing importance of by-products and coproducts. These changes could include either increased farmer ownership of gins or increased toll ginning, so that farmers can directly sell the seed. These changes will proceed most rapidly where farmers have reasonable organizational capacity and independent access to finance. In this respect, African sectors could be very slow to adopt these changes, and African farmers could thus be slow to benefit from these emerging markets.

# Chapter 9: Cotton Research
## *Duncan Boughton and Colin Poulton*

Farm-level productivity gains are critically important for Africa's cotton sectors to improve their international competitiveness and contribute more effectively to raising rural incomes and reducing poverty. Figure 10.1 shows that while world average yields for rainfed cotton production have increased by more than 150 percent since 1980, yields in Africa over the same period have risen by much less and have been essentially flat in West and Central Africa (WCA) since 1985. Lagging yields are an indicator of possible weaknesses in the technology development and delivery value chain that, if carefully diagnosed and corrected, signal a potential opportunity for raising competitiveness in the future. While a complete diagnostic is beyond the scope of the current study, this chapter provides an initial assessment of the current state of research and its contribution to competitiveness based on the case study countries.

This chapter also shows a high level of variation in productivity among some African cotton farmers. The most productive farmers achieve cotton yields close to or above the world rainfed average, suggesting that a major cause of low average yields in Africa is the inability of the majority of farmers to access or use existing technology packages effectively. Research is generally thought of as pushing out the production frontier, but the fact cannot be ignored that many cotton farmers in Africa are achieving yields far short of the existing frontier. At a minimum, socioeconomic research can help improve the understanding of the reasons for the yield performance gap, even if researchers may not be able to solve the underlying problems. For example, if closing the yield gap using existing technology requires additional investments in farmers' asset bases (biological, physical, and human capital), then additional resources beyond technical research, and indeed, from outside the cotton sector, may be needed to achieve high yields. Improved technology, such as Bt cotton seed or weedicides, as part of an integrated pest management strategy may reduce some management requirements for asset-poor farmers, but may also increase the farmers' perception of risk from higher costs for purchased input. A first step toward defining the role of research in improving yields will be to understand the causes of low productivity in specific agro-ecologies and farm types to determine the mix of investments necessary. For the most asset-poor cotton farmers, an important contribution of research may be to identify more profitable alternative crops or off-farm activities for them to take up instead of growing cotton.

## Cotton Research Organization and Performance

The organization and financing of cotton research is highly dependent on the historical context at both regional and country levels. In Francophone WCA, cotton research reflects a long history of investment by the former French cotton research institute IRCT (Institut de Recherches sur le Coton et les Textiles), with cofinancing at the country level by the former cotton parastatal CFDT (Compagnie Française pour le Développement des Fibres Textiles), and good information exchange among researchers in the region. Cofinancing of cotton research by cotton companies in Francophone countries continued after independence, often supplemented by development loans or grants, because cotton was seen as an "engine of development" that enabled the capitalization of smallholder farming. The persistence of the parastatal monopoly model of cotton sector management in WCA countries clearly contributed to organizational stability for cotton research when compared with East and Southern Africa (ESA), in addition to providing farmers with equipment and extension advice so they could adopt research recommendations. Most cotton research programs in WCA continue to be organized along the lines of individual disciplines, with variety, pest control, and agronomy subprograms. The research programs in Burkina Faso and Mali are the only ones in the study sample that appear to integrate a socioeconomic component on a systematic basis.

By contrast, the United Kingdom focused its colonial era research investments in regional commodity research programs (Beintema and Stads 2006), but cotton was not among them. Instead, individual colonies were left to finance cotton research through industry levies in those colonies where cotton was an important crop for the economy. This inevitably led to uneven performance across countries. However, Zimbabwe and Uganda in particular have strong traditions of cotton research.

With regard to human and financial resources, cotton research programs in WCA case study countries range from 9 to 25 research staff with annual budgets varying from US$300,000 to US$500,000 per year. Similar to publicly funded agricultural research in general, researchers often complain that high fixed costs, in part related to high support staff levels, often do not leave sufficient operational resources for field research activities.[64] Research programs in ESA case study countries are smaller than in WCA in absolute terms (ranging from 3 to 11 research staff). Although these resources may not be smaller in proportion to the production of the national cotton sector, critical mass can be important in sustaining a dynamic research program. As in WCA, there is great variability among countries in the ESA region, with Zimbabwe having the most and Mozambique having the fewest researchers.

Internal and external links are important to the effectiveness and impact of agricultural research. Internal links refer to relationships among national stakeholders, particularly institutional links between research and extension, as well as direct contact between researchers and company and farmer clients (considered in more detail later in this chapter). Virtually all countries have been trending toward lower extension worker–farmer ratios, with a high share of extension workers' time devoted to input delivery and credit management (chapter 6). External links, such as regional research networks, are important for leveraging the impact of country-level research resources. These links were strong in Francophone countries in previous decades, but have since weakened, while they are nonexistent or weak in ESA. The Association for Strengthening Agricultural Research in Eastern and Central Africa has no cotton research network, while the West and Central African Council for Agricultural Research and Development has recognized the need to rejuvenate cotton in its strategic plan. Reports of the International Cotton Advisory Council tri-annual international research meetings indicate that participation by African researchers is very limited.

As one indicator of research performance, table 9.1 shows the cumulative number of new seed varieties released by the research systems in this book's focus countries over the period 1985–2005.[65] WCA countries made many varietal introductions in the 1980s and early 1990s. However, the rate of introduction of new cultivars has slowed since 1995 in all three WCA countries for which data are shown. This evidence is consistent with the general loss of efficiency observed within WCA cotton sectors over the same period as the CFA franc

**Table 9.1 Number of Varietal Releases in Study Countries**

| Country | Cumulative number of varietal releases | | |
|---|---|---|---|
| | 2000–05 | 1995–2005 | 1985–2005 |
| Burkina Faso | 1 | 1 | 10 |
| Cameroon | 1 | 2 | 7 |
| Mali | 4 | 6 | 27 |
| Mozambique | 2[a] | 2 | 2 |
| Tanzania | 0 | 0 | 1 |
| Uganda | 0 | 3 | 6 |
| Zambia | 1[b] | 1 | 3 |
| Zimbabwe | 2 | at least 4 | at least 8 |

*Source:* Country case study reports.

a. These two varieties were introduced with support from the Mozambique Cotton Institute in collaboration with ginning companies, but they were not officially released by the research system.

b. In addition to the one variety released, up to six promising varieties are in the pipeline.

devaluation first eased financial pressure on the sectors and the expanding size of the sectors made them increasingly susceptible to political pressures. Meanwhile, within ESA, Uganda and Zimbabwe have produced a steady stream of varietal releases, although no new varieties have been released in Uganda in the past five years. By contrast, cotton research in Mozambique, Tanzania, and Zambia has a poor record over the past 20 years, as measured by varieties officially released, although there is some evidence of renewed vitality since the early 2000s.

In addition to varietal development, WCA countries invested considerable effort in the development of integrated pest management (IPM) and pest scouting methods to reduce costs and pest resistance build-up. Unlike varieties, where companies can enforce adoption, uptake of IPM methods by farmers has tended to be slow despite potential increased returns to adoption. Other crop management innovations have met with greater acceptance by farmers. The use of organic fertilizer in Mali and the rapid uptake of herbicides and direct sowing in Cameroon are examples of emerging success stories. The variable adoption rates of different technologies indicate the need for greater integration of social scientists in on-farm testing and early adoption studies.

## Articulating Research Demand: Insights from the Typology?

Despite historical influences on the performance of research in Africa, the positive engagement of the private sector in cotton research in other parts of the world (for example, Australia and Brazil) suggests that the ginning industry could also influence performance in Africa if ginners were encouraged to contribute to research management, perhaps as part of a wider public-private partnership for cotton sector research. Under such circumstances, ginners could do the following:

- Fund research through direct contributions. Alternatively, ginners might be the conduit for levies on cotton production activities to support research.

- Work to attract complementary donor funding for research efforts.

- Be involved in setting research priorities.

- Monitor research performance and demand an accounting of how research funds have been used.

- Given that cotton research is often underfunded and suffers from the weak management characteristic of many African public sector organizations, big improvements in research performance might be expected if ginners were to be so involved.[66]

The main hypothesis derived from the typology of cotton sectors is that smaller numbers of ginners (within concentrated sectors or monopoly systems) should find it easier than large numbers of players in a competitive sector to agree on their relative contributions to the funding of research and should have stronger incentives to monitor research performance in the detail and for the length of time required to begin to see improvements in the efficiency with which funds are used.[67] Thus, concentrated sectors and monopoly systems could be expected to perform better than competitive ones in research.[68] However, competitive sectors with many players could overcome this weakness through an effectively managed research fund to which all are required to contribute.

In practice, the link between current cotton sector structure and research performance is weak (table 9.1). A key reason is that, despite much investment aimed at improving the performance of public research institutions, overhauling the systems of incentives and accountability for scientists has not been possible. Most research programs remain firmly within the public sector, with researchers appointed through normal public sector channels, paid on public sector salary scales, and accountable primarily to national public sector research managers. Some signs of renewed research vitality are observed where these conditions have been slightly relaxed. However, there is still a long way to go in most cases. At the same time, governments have been slow to allow private ginners to contribute to research management, even where they have been allowed to assume responsibility for most other aspects of national cotton production and marketing.

In Tanzania and Mozambique, cotton companies have contributed to research funding through levies on seed cotton purchased or on lint exported. In Mozambique, there has been periodic pressure from some companies to reduce the value of the levy, which has contributed to a sense of instability and insecurity within the research program; such instability is clearly detrimental to performance. Meanwhile, other companies have opted to work directly with researchers (both national and international), including on testing new seed varieties. The Mozambican company Lomaco was the main example during the late 1990s, but its initiative was abruptly ended when parent company Lonrho ceased its involvement in Africa in the late 1990s. In a promising development, the Mozambique Cotton Institute brought together researchers (government and university) and interested companies to multiply and disseminate seed varieties developed for similar agro-ecological conditions in other countries, as well as to test improved crop management practices. Stakeholders meet one day a year to report on and plan research activities. This low transaction cost approach, plus company input into the subsequent work, seems to have convinced companies that it is an initiative worth their participation.

In Tanzania, despite the existence of the Tanzania Cotton Association (TCA), ginners' participation in the management of the sector-wide Cotton Development Fund, and several years of uncertainty over the status and quality of the new UK91 seed, ginners have not organized to demand greater accountability of the cotton research system to industry stakeholders. In 2006, TCA set up a Crop Development Committee to consider all matters related to production enhancement (presumably including research). However, as of early 2007 the committee had not held its first meeting. This is perhaps symptomatic of the incentive problem (free-riding) in a competitive market structure. Nevertheless, the fact that much research funding came through Cotton Development Fund levies for a few years may have raised expectations about performance. In 2006, the government committed once again to pay for research directly as a way of reducing levies within the sector. As a result, the flow of funds became too unreliable to sustain serious research activity. Breaking the stakeholder-funding link was a step backward.

In Zambia since 1999, research has been the responsibility of the Cotton Development Trust (CDT), which is intended to be semi-autonomous to encourage greater participation by other stakeholders, including farmers and ginners. Ginners were skeptical of CDT at its inception, given the moribund nature of the state research apparatus that it replaced and the fact that many of the staff were the same. To counter this, CDT sought to build relationships with ginners and farmers, including involving them in research priority setting and farm trials. However, ginners resisted the notion of a levy to fund research within the sector until they could see that funds might be used effectively. This highlights the point that ginners expect to monitor research performance and to see an accounting of how research funds are used in return for their contributions to funding the research. As of mid-2008, little active collaboration exists between the major ginners and CDT.

Within ESA the greatest involvement of the private sector in cotton research has occurred in Zimbabwe. As noted above, Zimbabwe has a strong tradition of public sector cotton research conducted in collaboration with the private sector. This approach dates back to an era when the Commercial Farmers' Union advocated strongly for public investment to support the interests and competitiveness of the country's white commercial farmers. After independence, close links were maintained between the parastatal Cotton Marketing Board (CMB) and the Cotton Research Institute (CRI). Following liberalization, Quton Seed Company, a wholly owned subsidiary of CMB's privatized successor Cottco, was granted monopoly rights to multiply and sell to the whole sector seed varieties produced by CRI, in exchange for which Quton made royalty payments to CRI. When state funding for CRI collapsed after 2001, these royalty payments became increasingly important, albeit inadequate to sustain a dynamic program. Quton, therefore, began to invest in its own research capability, which is now reckoned to surpass that of CRI. As of early 2007, Quton was preparing to release the first products of its in-house research activity and there was also talk of Quton taking over the assets of CRI. Quton's investment first in CRI and later in its own research capacity reflects the strong recognition within Zimbabwe of the importance of productivity-enhancing research. However, it is no coincidence that it has occurred in a concentrated sector where Cottco, as the largest firm, is also the biggest beneficiary of national research outputs.

Historically, the parastatals in WCA were heavily involved in funding cotton research. One consequence of this involvement was the focus within breeding programs on varieties with higher ginning outturn ratios (GORs) that lead to improved ginning margins, rather than on farm yield. In theory, higher ginning margins can be passed on to farmers, but it is not clear that farmers would have made this their breeding priority had they been consulted at the time. Also,

GOR is less visible to farmers than it is to ginners, and is less visible to farmers than is farm yield. Both these characteristics suggest that, especially where ginners have some market power, farmers will benefit more from increases in farm yields than from increases in GOR. Since 2000, the main organizational change in WCA has been a trend toward joint financing of research by all stakeholders through the inter-professional organizations set up in response to financial crises.

Research systems in WCA and ESA will increasingly have to respond to demand from international traders and spinners to take into account a more complex set of fiber quality parameters in variety selection (chapter 7). Increasing the number of parameters can dramatically complicate breeding programs, suggesting even greater need for increased resources for varietal research in these countries.[69] Also, because economies of scale in research may be high where several countries share common agro-ecological conditions, these more complex demands on breeders provide additional arguments for strengthening regional collaboration in these programs.

## Conclusions

Overall, research in African cotton systems is underperforming, as evidenced primarily by the slow rate of yield growth and technology adoption in Sub-Saharan Africa relative to other major cotton growing areas of the world. Reasons for this poor performance include the limited resources invested in cotton research, lack of incentives and accountability for publicly funded researchers, weak links among researchers in different countries, and the limited responsiveness of cotton research programs to ginners and farmers. Governments have been slow to allow private ginners to contribute to research management, while farmer associations are rarely strong enough, especially in ESA, to articulate firm demands for high quality research effort.

The reluctance of the public sector to relinquish or share control over cotton research, plus the long lags inherent in agricultural research, mean that the current structure of the ginning industry exerts a weaker influence over research performance than has been hypothesized in this analysis, and that history exerts a relatively greater influence. However, recent experience in Mozambique, Zambia, and Zimbabwe, combined with the trend in WCA toward joint financing of research by all stakeholders through the inter-professional organizations, suggests that concentrated sectors and monopoly systems may, over time, begin to perform better than competitive ones with regard to research.

Irrespective of sector type, a high policy priority should be to move toward greater involvement of both ginners and farmers in research management. The precise model, which can involve a greater or lesser role for state agents, can be worked out locally. However, ideally it should

- include stakeholder involvement in setting research priorities, monitoring research performance, and accounting for how research funds have been used;

- reward researchers who achieve high performance and meet stakeholders' expectations;

- include majority funding of research effort by the industry through direct contributions or levies; stakeholders should also work together to attract outside funding to complement internal resources, and such collaboration will be facilitated by the confidence that comes from working together to set priorities and monitor performance;

- allow stakeholders, through whatever management regime is put in place, to play a key role in researchers' evaluation, pay, and promotion (or termination), rather than relying on public sector practices and scales; and

- feature more active regional information sharing, strategic priority setting, and eventually financing across countries in similar agro-ecological zones.

With improved funding, human resources, and management, cotton research could yet play a vital role in helping Africa's cotton sectors improve competitiveness. Specific opportunities for technological progress include the identification of well-adapted varieties with stacked genes incorporating Bt and herbicide resistance, integrated soil and water management strategies to reduce the impact of rainfall variability and adapt to rapidly increasing costs of inorganic fertilizers, and improved integrated pest management packages (including herbicides for farmers without animal traction). The extent to which these technologies improve the productivity of farmers with different asset bases needs to be carefully analyzed to avoid a one-size-fits-all approach and to identify complementary investments that may be needed to enable farmers to use technology effectively.

Finally, while renewed attention to research funding and management are essential if African cotton sectors are to shift their productivity frontier and remain competitive in a dynamic world, attention must also be give to assessing the potential to raise the productivity of farmers who are currently well below the frontier. As the next chapter argues, these efforts are likely to be long term and to involve the promotion of asset accumulation (especially animal traction), along with the design of systems that are able to supply credit even to poorer farming households.

# SECTION IV
# COMPARATIVE ANALYSIS: OUTCOMES

# Chapter 10: Yields and Returns to Farmers
## Colin Poulton, Patrick Labaste, and Duncan Boughton

Profitability of cotton cultivation at the farm level is one of the key expected outcomes of a viable and sustainable cotton sector and as such a fundamental indicator of sector performance. Profitability in turn depends on several main variables: productivity at field level, as reflected in seed cotton yields; prices paid to farmers for their seed cotton (already analyzed in chapter 6), eventually including quality premiums; and costs incurred by farmers in the production process. Profitability should also include the valorization of cotton seed, even though, as discussed in chapter 8, African farmers generally do not receive a specific payment for the seeds.

As pointed out in the previous chapter, farm-level world average yields for rainfed cotton have increased sharply and steadily since 1980, while average yields in Sub-Saharan Africa (SSA) have not kept pace during the same period. This chapter first examines the data on yields across countries and sector types; it then assesses returns to farmers based on crop budgets evaluated according to categories of farmers used in sample surveys carried out using the participatory rural appraisal focus group methodology described in the annex to this chapter.[70]

## Cotton Yields

Cotton yields on farmer fields are an important determinant of sector competitiveness and its impact on poverty reduction. This section focuses on yield trends in global and regional contexts, factors underlying those trends, and their relationship to different ways of organizing the cotton sector. When comparing across sector types, it is helpful to consider average yields and yield trends. The typology in chapter 4 suggests that average yields will be strongly influenced by performance on input provision and extension (expected to be best in monopolies and concentrated sectors), but will also depend heavily on past investments, especially in research.

### COTTON YIELDS IN A GLOBAL CONTEXT

World cotton yields expressed in lint equivalent have increased from an average of 300 kg per hectare (ha) in the early 1960s to more than 700 kg/ha in 2005, equivalent to a 1.9 percent annual increase (figure 10.1). Yields in West and Central Africa (WCA) increased at an annual rate of 5 percent until the early 1980s, reaching about 450 kg/ha at that time, mainly in response to the introduction of fertilizer (a characteristic in all cotton producing WCA countries). Yields in this region, however, have stagnated or even declined slowly over the past 20 years.[71] Yields in East and Southern Africa (ESA) have been increasing at the same rate as the world average since the early 1970s, albeit from a very low base. Average yields in ESA as of mid-2008 are slightly more than one-half the WCA average and one-third the world average.

To understand the factors underlying long-term yield trends, a distinction must first be made between irrigated and rainfed production systems. Africa's cotton is almost entirely rainfed, while world average yields reflect the fact that 55 percent of cotton is produced under irrigated conditions. Though average yields under irrigation are much higher than under rainfed conditions, worldwide yield growth has been much greater in the latter: average yield in rainfed

**Figure 10.1 Cotton Yield Trends in World, WCA and ESA (1970-2005)**

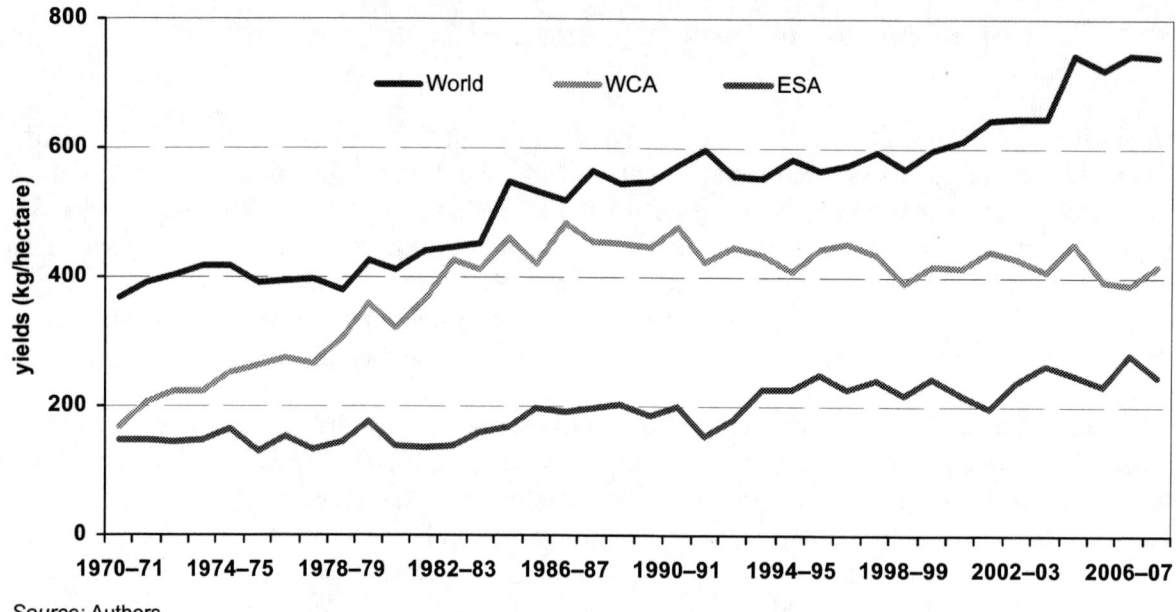

*Source:* Authors.

cultivation more than doubled between 1980 and 2005, growing 3.9 percent per year, while yield in irrigated systems increased by only 60 percent, or 1.8 percent per year (figure 10.2). Yet Africa, where nearly all production is rainfed, has not seen this kind of growth: ESA yields have risen only about 2.1 percent per year, while WCA yields have stagnated or declined. As a result, while WCA yields were well above world average rainfed yields in 1980/81, the rest of the world's rainfed cotton production systems have now surpassed WCA. What has been driving these divergent yield trends?

**Figure 10.2 Average Worldwide Yields of Irrigated and Rainfed Cotton**

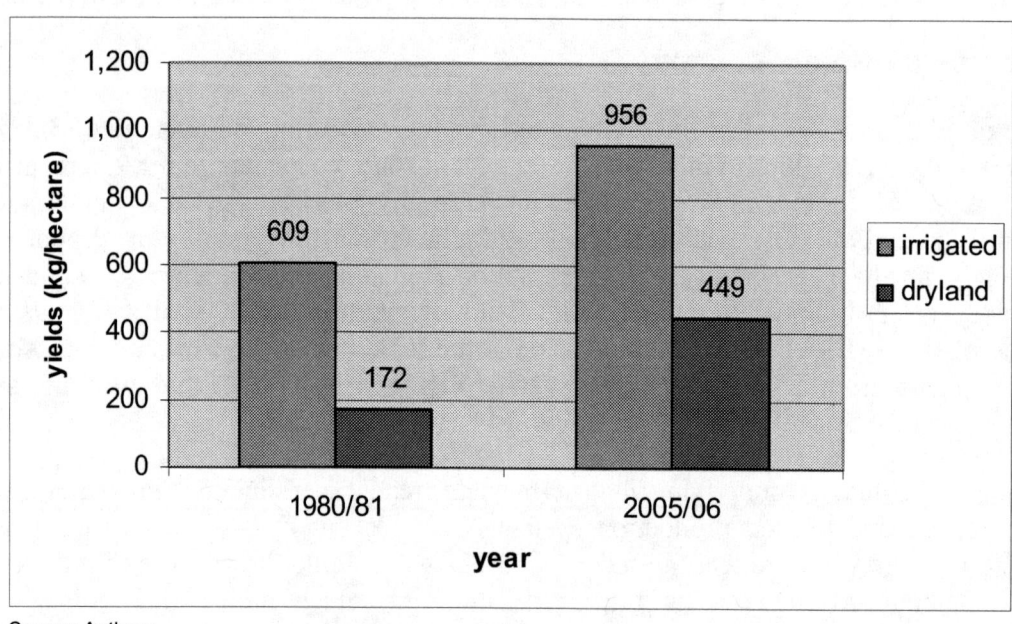

*Source:* Authors.

## EVOLUTION OF COTTON YIELDS IN WCA AND ESA

Declining cotton yields in WCA are commonly explained by (i) declining fertilizer use, partly resulting from a reallocation to food crops; (ii) a decline in the number of insecticide applications, leading to higher infestation levels; (iii) problems with seed quality, often quoted as a reason for declining yields in interviews with farmers in Mali in 2007; and (iv) degradation of soil fertility from continuous cultivation and insufficiently adapted fertilization formulas. These factors are discussed in more detail later in this chapter.

Country average yields over time mask variation within countries from spatial and socioeconomic factors. In Benin, for example, during the 2001/02 cropping season, seed cotton yields ranged from 810 kg/ha in the south to 1,318 kg/ha in the north, primarily as a result of the introduction of a more productive variety in the north. Spatial factors linked to climatic and soil conditions also contribute to large differences in yields between southern and northern Cameroon (670 kg/ha and 1,500 kg/ha, respectively). In both regions of the country, yields declined following the reduction of fertilizer subsidies. The role of socioeconomic differentiation can be illustrated by Mali, where farm yields in 2002/03 ranged from 1,090 kg/ha with manual cultivation to 1,259 kg/ha for farms equipped with their own animal traction. The role of socioeconomic differentiation is discussed later in this chapter.

Figure 10.3 illustrates differences in cotton yields (lint equivalent) among the major ESA cotton producers, based on a 20-year average. ESA country averages are well below world and WCA levels, and vary by a factor of more than two between Mozambique (lowest) Zimbabwe (highest). Most cotton sectors in ESA are based on a low input–low output system. For example, outside Zimbabwe virtually no fertilizer is applied by ESA cotton farmers, who benefit from better soils than farmers in WCA. Countries in ESA generally show greater inter-annual yield variability from climatic events than countries in WCA.

The steady yield improvement in ESA is due to improved varieties and, in some countries, improved input use. Zambia has shown a steady (but slow) upward yield trend since

**Figure 10.3 Average Yields of Rainfed Cotton in ESA Countries, WCA, and World, 1994/95–2003/04**

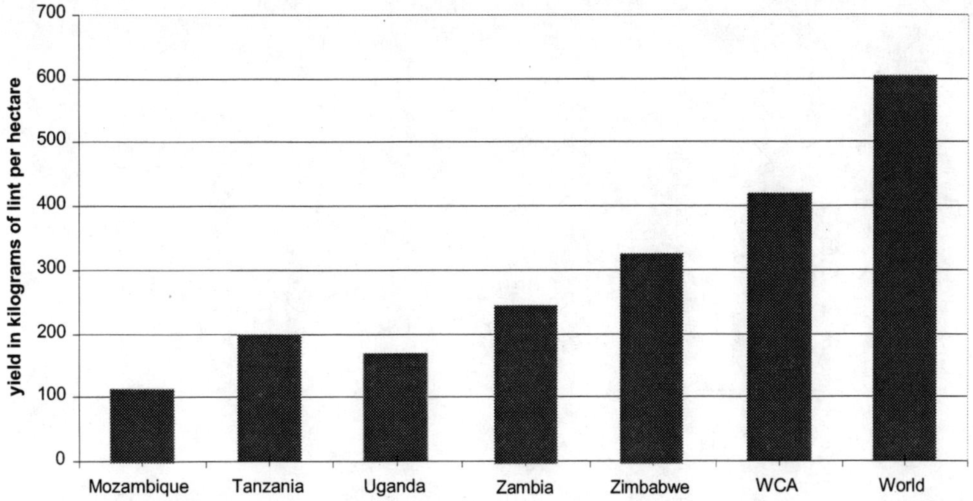

*Source*: International Cotton Advisory Committee.

liberalization; the input distribution and extension efforts of the two dominant companies are the main reasons. However, ESA has not seen rapid increases in productivity like those in China, India, and Pakistan, where genetically modified cotton varieties were introduced.

## YIELDS BY FARMER TYPE

To assess performance of different categories of farmers, focus group discussions were undertaken as part of this study in seven of the nine study countries (all except Benin and Cameroon). Respondents were asked to group farmers in their area according to volumes of production. Generally speaking, the groups can be thought of as large (group 1), medium (group 2), small (group 3), and very small (group 4). In some countries, respondents identified only three groups (large, medium, and small) although what constitutes each category varies by country context. Additional details about the methodology of this research, and its limitations, can be found in the annex to this chapter. A key insight emerging from the farmer group interviews concerns the variation in yields across groups within individual countries, which are at least as great as the variations across countries. These are shown in figure 10.4.

There are two main causes of this variation across groups within a country. First, differences in access to input are an important factor in ESA, but much less so in WCA. Largely as a result, the variation in yields across groups is less pronounced in WCA than in ESA. The average ratio of yields between the top and bottom groups in the three WCA cotton sectors is 1.65, compared with 6.4 in the five ESA sectors. A second reason is differences in ownership of assets, of which the most important are arguably oxen and plowing equipment. Households that own their own animals and plows can prepare their land as soon as the rains begin, thus permitting timely planting (a prerequisite for good yields) and the cultivation of larger areas of land. Larger producers also tend to have family labor or the working capital to hire labor in a timely fashion. The poorest farmers are often caught in a food insecurity trap, which causes them to hire out their labor for immediate cash income instead of working on their cotton plots.

**Figure 10.4 Variation in Yields across Farmer Groups**

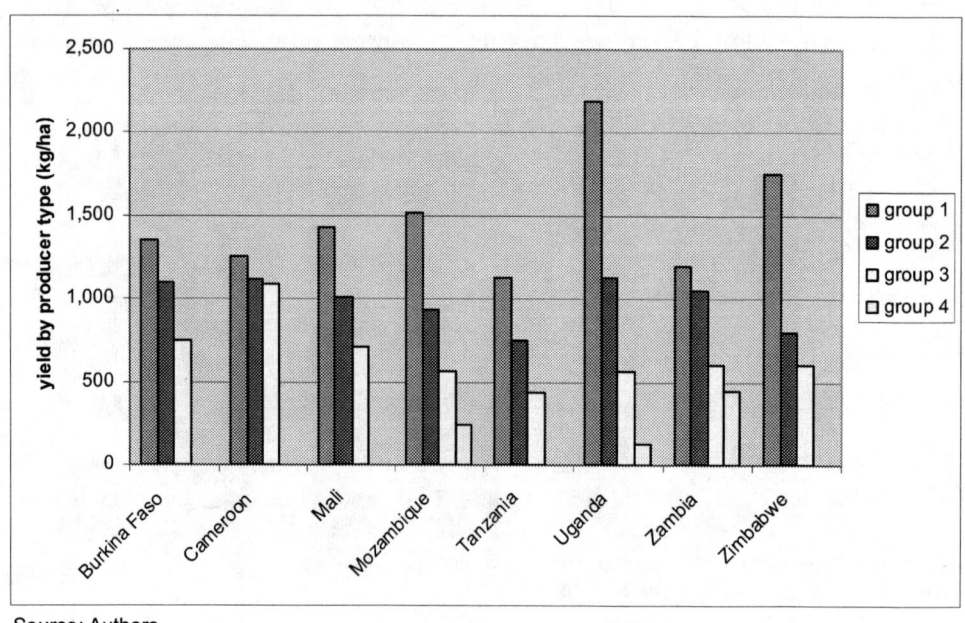

Source: Authors.

92

Yield levels and trends at regional and country levels are correlated with cotton sector organization. The national monopoly systems established in the Francophone countries of WCA delivered impressive and sustained yield growth over a period of three decades, from very low yields in the 1950s to well above the world rainfed average yield in the 1980s (around 1,200 to 1,400 kg/ha of seed cotton). This achievement was due to a reliable system for varietal development, input supply and credit, quality extension services, and logistical organization provided by the cotton companies. Since the mid-1980s, this trend has not been sustained and the productivity gap has started to widen. The system has not demonstrated a capacity to adapt to changing technical and economic circumstances, particularly with regard to making improved technical packages available to farmers.

In ESA countries, the trend has been slow but steady increases in yields from a low base, based on low input–low output production systems, reaching about half the world average for rainfed cotton today. Variation in yield performance among ESA countries is also correlated with sector organization. Yields are higher in the more concentrated systems (Zambia and Zimbabwe) than in the more competitive models (Tanzania and Uganda), which have found it very difficult to provide the services farmers need to raise their yields (figure 10.3).

## Returns to Farmers

The evolution and typology of cotton sectors presented in chapter 4 gives rise to an important question given the study objectives: from which type of system do farmers consistently benefit more, a competitive system that pays them a higher share of world lint prices, but is less effective at delivering support services that help them raise yields, or a more coordinated sector that delivers reasonable support services but a lower share of the world price? A corollary question is whether some types of farmers do better under one system and others under another. Because farmer welfare depends on several factors, and because no one sector is expected to perform best on all these factors, the typology delivers no clear direction on how farmers will fare under different sector types. Yet answers to these questions are crucial to informing the design of reforms that improve competitiveness and accelerate poverty reduction.

Factors beyond company services also influence returns to farmers and their costs of production; the analytical challenge is to disentangle these from the impacts of the type of sector organization. In particular, the discussion below notes the impact of

- historical investment by cotton sector stakeholders, especially investments by companies to promote, and by farmers to adopt, animal traction; and

- differences in soil fertility across cotton growing regions and countries.

This section uses crop budget data to generate two key indicators: returns to family labor and returns to total labor (including hired). In previous work (for example, Poulton et al. 2004), crop budgets were disaggregated by types of farmer in each country. The types or groups of farmers are those identified by focus group discussions in seven of the sample countries. In Cameroon, where no focus group discussions were undertaken, data from SODECOTON monitoring

**Figure 10.5 Proportion of Cotton Farmers by Farmer Group**

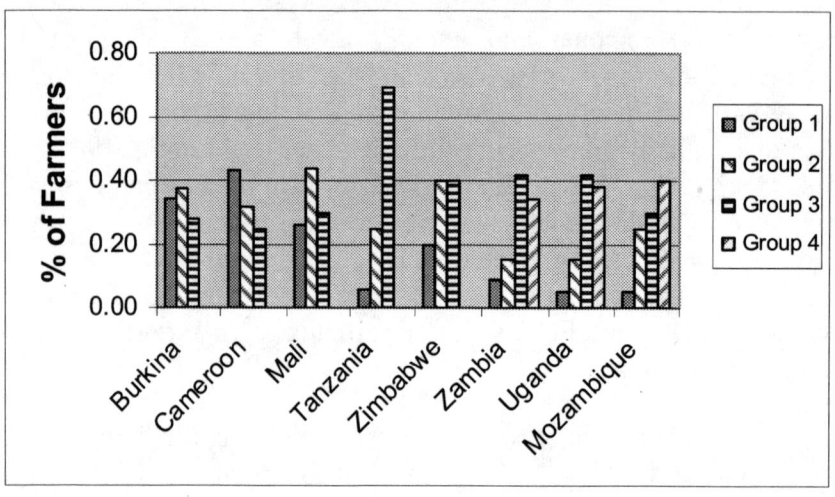

*Source:* Authors.

surveys were used. Figure 10.5 shows estimates of the proportion of cotton producing households by group in each country.

**IMPACT OF INPUT ON CROP BUDGETS**

Table 10.1 presents summary crop budgets by farmer type. The focus is on the use of labor and other inputs and the two key farm-level indicators: returns to family labor and returns to total labor. Table 10.1 also presents data on costs of seed cotton production across groups and countries. The most striking finding is the increase in unit costs of production as one moves from the top producers (group 1) to the poor and less efficient ones (group 3 or 4).

According to the focus group informants, average labor input per hectare of cotton production is 40 percent higher in ESA than in WCA.[72] The main reason is the greater penetration of animal traction technology within WCA, where it is used by nearly all farmers for land preparation, and by many for weeding. Efforts are also being made to promote the use of labor-saving herbicides in WCA, whereas in ESA only the top two groups in Mozambique recorded any use of herbicides.[73] Thus, weeding is the single largest contributor to this labor-use differential between the two regions.

Comparing higher and lower producing groups, total labor input falls with production level. Smaller producers require less labor for harvesting, tend to weed fewer times, and require less labor (if any at all) for fertilizer application and spraying. However, smaller producers (group 3 in Mozambique, Tanzania, and Zambia; all group 4s) often do not have access to animal traction, even for land preparation, so may have to use hand hoes, which is much more labor intensive. Alternatively, they have to hire plowing services, which raises their expenditure on hired services above that of larger producers (see Burkina Faso group 3, for instance).

Focus group discussions in Mali and Burkina Faso did not distinguish between family and hired labor input, so in table 10.1 all labor input in the WCA countries is considered to be family labor. While this may not be entirely true, average family sizes are much larger in WCA than in ESA, and it is understood that most labor tasks on WCA cotton farms are performed by family members. By contrast, the top producer groups in ESA rely heavily on hired labor, which accounts for over 70 percent of total labor input for group 1 in Mozambique, Tanzania, Zambia, and Zimbabwe (and also in Zambia group 2). Smaller producers in ESA countries are more reliant on family labor.

As discussed in chapter 6, given the high input costs associated with cotton production and the difficulties that African smallholder households have in affording such inputs, one of the main strengths of coordinated cotton systems is their ability to provide producers with access to input on credit. Participants in focus group discussions in Burkina Faso and Mali insisted that all groups use the same quantity of fertilizers per hectare. This response may have been influenced by the presence of the local extension officer at the discussions. Yet all cotton farmers are entitled to receive a similar quantity of fertilizers (per hectare of cotton cultivated) on credit, so fairly uniform usage is credible. By contrast, fertilizer use is highly skewed in Uganda and Zimbabwe, the only two countries in ESA where any inorganic fertilizer is used on cotton. Despite a 50 percent subsidy in Uganda, only group 1 farmers reported using fertilizers. According to participants in the focus group discussions, fertilizer use by the top group in Zimbabwe is higher than in any other country, including WCA. However, in part as a result of nationwide fertilizer shortages in Zimbabwe in 2005/06, fertilizer use by group 2 is much less than by group 1, while group 3 is not considered creditworthy enough to receive credit for fertilizer, even under normal circumstances.

The reasons that inorganic fertilizer is not used in Mozambique, Tanzania, or Zambia vary by country.[74] In Tanzania, the highly competitive sectoral structure makes recovery of input loans impossible, and the passbook system (see chapter 6) is not designed to enable access to inorganic fertilizers, which are much more expensive. In Mozambique and Zambia (and, for the time being, perhaps also Tanzania), it is questionable whether farmers really need inorganic fertilizers, given existing soil fertility levels and the generally moderate fertilizer response of cotton. Particularly in Mozambique, the relatively high levels of soil fertility are a major advantage for the cotton sector. It is inconceivable that the yields claimed by group 1 farmers in Mozambique could be achieved by farmers in other countries without the use of either inorganic fertilizer or manure. At the other end of the spectrum, producers in WCA and Zimbabwe grow cotton on less productive soils.

There is less variation in the provision and use of plant protection chemicals across countries and groups than there is for fertilizers. Nevertheless, usage is more uniform across groups in WCA countries than in ESA.

Table 10.1 also shows the ratio of input costs to gross revenues for each producer type and sector. This ratio is an important indicator of the risk entailed in cotton production. A cross-country comparison shows that the most important determinant of the ratio of input costs to gross revenues is the quantity of inorganic fertilizer used. The ratio is thus higher in WCA and Zimbabwe (which also has the highest number of pesticide sprays per season) than in Mozambique, Tanzania, Uganda, or Zambia.

## Table 10.1 Summary Crop Budgets by Farmer Type and Country

| Budget element | Burkina Faso | Cameroon[a] | Mali | Mozambique | Tanzania | Uganda | Zambia | Zimbabwe |
|---|---|---|---|---|---|---|---|---|
| **Group 1** | | | | | | | | |
| Yield (kg/ha) | 1,350 | 1,259 | 1,429 | 1,519 | 1,125 | 2,188 | 1,200 | 1,750 |
| Seed cotton price (US$/kg) | 0.33 | 0.32 | 0.32 | 0.21 | 0.28 | 0.25 | 0.25 | 0.31 |
| Gross revenue (US$/kg) | 441.45 | 399.10 | 452.99 | 322.03 | 314.06 | 547.00 | 300.00 | 542.50 |
| Cost of input (US$/ha) | 172.89 | 141.44 | 168.61 | 36.50 | 35.83 | 111.11 | 31.07 | 236.85 |
| Cost of hired services (US$/ha) | 28.52 | 48.83 | 33.69 | 22.25 | 54.18 | 72.22 | 17.06 | 32.98 |
| Cost of hired labor (US$/ha) | 0 | 0 | 0 | 136.70 | 122.90 | 116.27 | 150.71 | 65.10 |
| Gross margin, excluding labor (US$/ha) | 240.04 | 208.83 | 250.69 | 263.28 | 224.06 | 363.67 | 251.87 | 272.67 |
| Returns to all labor[b] (US$/day) | 2.38 | 1.37 | 2.63 | 1.36 | 1.60 | 2.76 | 1.73 | 2.19 |
| Returns to family labor (US$/day) | n.a. | n.a. | n.a. | 2.56 | 3.37 | 3.81 | 2.68 | 6.15 |
| Cost per kg[c] (US$) | 0.22 | 0.32 | 0.21 | 0.15 | 0.22 | 0.19 | 0.20 | 0.21 |
| Family labor input (days/ha) | 100.7 | 152 | 95.4 | 49.5 | 30.0 | 65.0 | 37.8 | 33.8 |
| Hired labor input (days/ha) | 0 | 0 | 0 | 144.8 | 110.0 | 67.0 | 108.0 | 90.7 |
| Net margin (US$/ha) | 140.34 | −1.86 | 156.24 | 91.93 | 71.16 | 137.55 | 56.16 | 173.82 |
| Input cost / gross revenue | 0.39 | 0.35 | 0.37 | 0.11 | 0.11 | 0.20 | 0.10 | 0.44 |
| **Group 2** | | | | | | | | |
| Yield (kg/ha) | 1,100 | 1,120 | 1,011 | 935 | 750 | 1,125 | 1,050 | 800 |
| Seed cotton price (US$/kg) | 0.33 | 0.32 | 0.32 | 0.21 | 0.26 | 0.25 | 0.25 | 0.29 |
| Gross revenue (US$/kg) | 359.70 | 355.04 | 320.49 | 198.22 | 196.88 | 281.25 | 262.50 | 232.00 |
| Cost of input (US$/ha) | 164.89 | 132.76 | 159.58 | 36.00 | 18.00 | 8.33 | 31.07 | 90.08 |
| Cost of hired services (US$/ha) | 34.55 | 15.91 | 31.30 | 4.76 | 40.83 | 71.11 | 39.36 | 25.38 |
| Cost of hired labor (US$/ha) | 0 | 0 | 0 | 116.80 | 42.71 | 62.50 | 109.52 | 35.85 |
| Gross margin, excluding labor (US$/ha) | 160.26 | 206.37 | 129.61 | 157.46 | 138.05 | 201.81 | 192.07 | 116.55 |
| Returns to all labor[b] (US$/day) | 1.78 | 1.36 | 1.64 | 0.66 | 1.16 | 2.15 | 1.44 | 1.08 |
| Returns to family labor (US$/day) | n.a. | n.a. | n.a. | 0.28 | 1.19 | 2.49 | 2.65 | 1.31 |
| Cost per kg[c] (US$) | 0.26 | 0.32 | 0.27 | 0.28 | 0.24 | 0.21 | 0.21 | 0.27 |
| Family labor input (days/ha) | 90.2 | 152 | 79.1 | 143.0 | 80.0 | 56.0 | 31.2 | 61.4 |
| Hired labor input (days/ha) | 0 | 0 | 0 | 95.8 | 39.5 | 38.0 | 102.0 | 46.1 |
| Net margin (US$/ha) | 70.95 | −4.32 | 51.29 | −59.44 | 15.335 | 44.67 | 45.41 | 19.33 |
| Input cost / gross revenue | 0.46 | 0.37 | 0.50 | 0.18 | 0.09 | 0.03 | 0.12 | 0.39 |

*(continued)*

| Budget element | Burkina Faso | Cameroon[a] | Mali | Mozambique | Tanzania | Uganda | Zambia | Zimbabwe |
|---|---|---|---|---|---|---|---|---|
| **Group 3** | | | | | | | | |
| Yield (kg/ha) | 750 | 1,090 | 711 | 438 | 600 | 600 | 563 | 565 |
| Seed cotton price (US$/kg) | 0.33 | 0.32 | 0.32 | 0.24 | 0.24 | 0.25 | 0.25 | 0.21 |
| Gross revenue (US$/kg) | 245.25 | 345.53 | 225.39 | 103.91 | 144.00 | 150.00 | 140.75 | 119.78 |
| Cost of input (US$/ha) | 156.89 | 141.44 | 146.04 | 5.50 | 48.55 | 20.41 | 8.33 | 13.50 |
| Cost of hired services (US$/ha) | 75.87 | 11.61 | 23.10 | 1.04 | 0 | 9.93 | 71.11 | 0.80 |
| Cost of hired labor (US$/ha) | 0 | 0 | 0 | 0 | 0 | 28.57 | 5.21 | 6.70 |
| Gross margin, excluding labor (US$/ha) | 12.49 | 192.48 | 56.25 | 97.37 | 95.45 | 119.66 | 61.31 | 105.48 |
| Returns to all labor[b] (US$/day) | 0.17 | 1.27 | 0.99 | 1.02 | 0.90 | 0.78 | 0.72 | 0.49 |
| Returns to family labor (US$/day) | n.a. | n.a. | n.a. | 1.02 | 0.90 | 0.72 | 0.74 | 0.49 |
| Cost per kg[c] (US$) | 0.41 | 0.33 | 0.32 | 0.23 | 0.26 | 0.35 | 0.38 | 0.29 |
| Family labor input (days/ha) | 75.3 | 152.0 | 57.0 | 95.0 | 106.3 | 126.0 | 76.0 | 200.8 |
| Hired labor input (days/ha) | 0 | 0 | 0 | 0 | 0 | 27.0 | 9.0 | 13.1 |
| Net margin (US$/ha) | −62.06 | −18.21 | −0.19 | 2.37 | −10.80 | −58.91 | −72.34 | −41.78 |
| Input cost / gross revenue | 0.64 | 0.41 | 0.65 | 0.05 | 0.34 | 0.14 | 0.06 | 0.11 |
| **Group 4** | | | | | | | | |
| Yield (kg/ha) | | | | 240 | | 125 | 450 | |
| Seed cotton price (US$/kg) | | | | 0.21 | | 0.25 | 0.25 | |
| Gross revenue (US$/kg) | | | | 50.88 | | 31.25 | 112.50 | |
| Cost of input (US$/ha) | | | | 13.50 | | 5.56 | 16.86 | |
| Cost of hired services (US$/ha) | | | | 2.00 | | 0 | 10.33 | |
| Cost of hired labor (US$/ha) | | | | 5.30 | | 0 | 0 | |
| Gross margin, excluding labor (US$/ha) | | | | 35.38 | | 25.69 | 85.31 | |
| Returns to all labor[b] (US$/day) | | | | 0.24 | | 0.28 | 0.70 | |
| Returns to family labor (US$/day) | | | | 0.24 | | 0.28 | 0.70 | |
| Cost per kg[c] (US$) | | | | 0.45 | | 1.27 | 0.38 | |
| Family labor input (days/ha) | | | | 123.7 | | 91.0 | 122.5 | |
| Hired labor input (days/ha) | | | | 25 | | 0 | 0 | |
| Net margin (US$/ha) | | | | −56.51 | | −128.10 | −60.52 | |
| Input cost / gross revenue | | | | 0.27 | | 0.18 | 0.15 | |

*Source:* Authors.

*Note:* n.a. = Not applicable.

a. Because no focus groups were undertaken, Cameroon data are from monitoring surveys by SODECOTON. Labor data are not disaggregated by farmer type and are generally considered too high.

b. Calculations of returns to all labor assume no payment for hired labor.

c. For calculations of cost per kg, all labor input is costed at an average casual labor wage.

## RETURNS TO FARMERS

Two key indicators at farm level were used to evaluate returns to farmers: weighted average return to family labor, and return to all labor. First, however, farm profits per ha by farmer group are examined; profit data provide important insights into the potential poverty reduction effects of cotton cultivation.

After valuing family labor at the going casual wage rate in rural areas, group 1 households make a profit in all countries, while group 2 households make a profit in all countries except Cameroon and Mozambique (figure 10.6). In both Cameroon and Mozambique, apparent high labor requirements cause the loss. In Tanzania and Zimbabwe, the profit achieved by group 2 households means that, under current conditions, cotton production can only make a modest contribution to household income and poverty reduction objectives.

Tanzania is the only country where group 3 households make a profit, as defined here. This result means that the household obtains a higher income from applying its own labor to its cotton plot than it could from selling the same quantity of labor at the assumed casual wage labor rate. In Mali, the return to labor achieved by group 3 households is identical to the assumed casual wage labor rate.

The stark finding from these figures is that between 25 percent (Burkina Faso) and 75 percent (Mozambique, Uganda, and Zambia)[75] or more of cotton producing households would be better off hiring out their labor than applying it to their own cotton plots. Why do they persist in producing cotton?

**Figure 10.6 Net Margins after all Costs (including Labor), US$/kg**

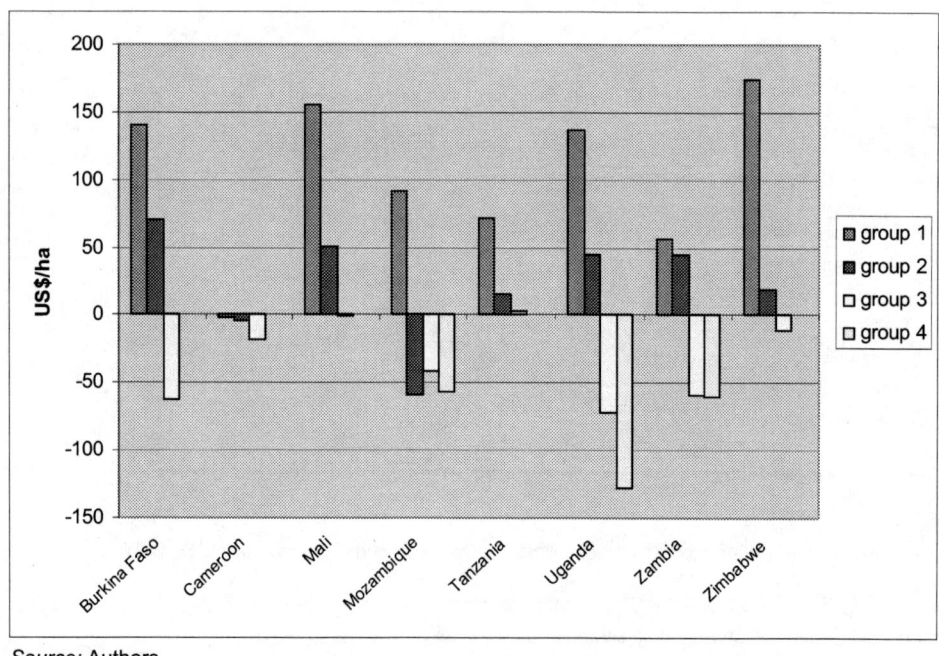

*Source:* Authors.

Two main answers emerged from the focus group discussions:

- First, many group 3 and 4 households prioritize the hiring out of their labor, then fit in cultivation of their own cotton farms when they are not working elsewhere. This is a major reason why these groups perform many of their critical cultural practices late and hence why they achieve such low yields.

- Focus group participants argued that cotton is the most remunerative cropping activity available in their areas. A critical factor here is the reliable market provided by cotton companies, which means that farmers can be sure of obtaining at least some cash income (a scarce commodity in group 3 and 4 households) from cotton production.

Weighted average returns to family labor and to all labor are presented in figure 10.7; the weights are the proportion of farmers by farmer group shown in figure 10.5. (See table 10.1 for the returns figures for each group). Returns to family labor and to all labor are identical for the WCA countries because all labor was recorded as family labor (and, in fact, very little hired labor is used). The WCA countries boast three of the four highest returns to family labor, and the three highest returns to all labor. This result is driven by the success of these systems in moving farmers into groups 1 and 2 over time. Zimbabwe delivers the highest return to family labor, again in part reflecting the efforts made first by its Cotton Marketing Board and then by Cottco and Cargill to support farmers with extension services and input access over a sustained period. All other ESA countries lie below the three WCA countries. Mozambique performs especially poorly, reflecting its small share of households in group 1 and the very low prices paid to farmers.

**Figure 10.7 Weighted Average Returns to Family Labor and All Labor in Study Countries**

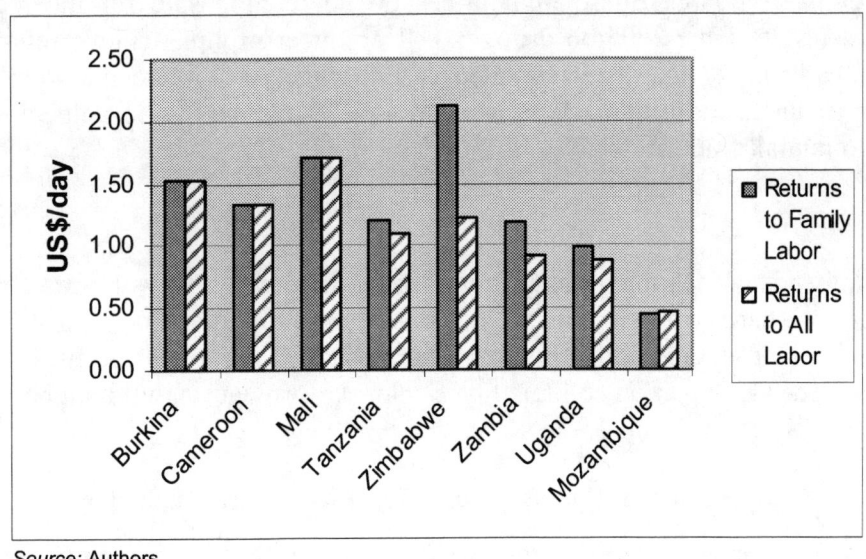

*Source:* Authors.

## ALTERNATIVE SCENARIOS

For varying reasons, the "base-case" budgets in Burkina Faso, Mali, and Tanzania represent a fairly optimistic scenario. Thus, in addition to the base-case budgets, the chapter also presents findings on returns under the following alternative scenarios:

- *Burkina Faso and Mali.* Large debts incurred by the cotton companies imply that seed cotton prices in 2005/06 were at an unsustainably high level. Budgets are thus reestimated assuming a seed cotton price of CFA franc 150 per kg (all other variables are held constant).

- *Tanzania.* A downward adjustment is applied to the seed cotton yields reported by the focus groups, to reflect national average yields in the 2005/06 drought season. These adjusted yields are then combined with actual seed cotton prices received in 2005/06. Harvest labor (but not other labor or cash inputs) is also adjusted to reflect the lower yields.

All other variables in the budgets are kept constant. Predictably, returns to labor fall for all groups. Critically, returns to labor become negative for group 3 in Burkina Faso, which means that cotton would be completely nonviable for this group, even if labor input into cotton did not compete at all with opportunities for off-farm labor. Recalculating the weighted returns to labor under these alternative scenarios, the figure for Burkina Faso becomes comparable to that achieved by Zimbabwe. This suggests that, under a more sustainable pricing regime, WCA sectors could still deliver returns to labor that are comparable to the best achieved in ESA.

The alternative scenario for Tanzania involved lower yields, but higher prices (the prices actually observed in 2005/06) and lower harvesting labor. The country report (Poulton and Maro 2007) noted that national yields in 2005/06 were only 58 percent of those recorded in 2004/05, so the base-case yield for each group was adjusted by this factor. Actual seed cotton prices paid in Tanzania in 2005/06 were extremely attractive—the highest outside of the subsidized WCA sectors and comparable to the sustainable prices just noted for Burkina Faso and Mali. However, even with lower harvest labor, these high prices do not compensate for the reduced yields experienced as a result of the 2005/06 drought. All producer groups achieve returns to family labor that are clearly lower than the estimated casual wage rate, while the weighted average returns now approximate those realized in Mozambique. This shows the vulnerability of ESA cotton farmers to rainfall fluctuations.

## CONCLUSIONS

The starting point when comparing across sectors is that there will always be some farmers who, by superior skill or hard work and asset accumulation (and perhaps higher starting asset endowment), do well, producing high yields and thereby achieving good returns. How well these top producers do depends in part on soil fertility. Sector performance should not be judged on the basis of the performance of this top group alone.

Companies (hence sectors) can contribute to overall farm-level performance by

- assisting with access to input through credit provision, thereby allowing more farmers to achieve high yields and allowing good farmers to expand production (at a given yield level);

- providing extension advice, thereby assisting more farmers to raise productivity;

- paying high prices, thereby raising returns for a given technical performance; and

- facilitating asset accumulation, especially animal traction. Asset accumulation can be assisted either passively (by assisting producers in obtaining good returns over a sustained period) or actively (by promoting uptake of the assets in question). Asset accumulation allows more producers to move into groups 1 and 2 over time.

An important lesson from this analysis is that assisting more farmers to move into groups 1 and 2 is critical for sector competitiveness (see data on costs of production in table 10.1) and for poverty reduction (see figure 10.4). Cotton sectors in WCA countries have assisted more farmers in rising to groups 1 and 2 over time through their promotion of animal traction, which allowed farmers to increase their cotton area and yields. The majority of farmers in WCA also enjoy access to input that only the top households in selected ESA sectors enjoy. At the same time, recent strong political pressures have contributed to high (though unsustainable) seed cotton prices for WCA producers. These broadly based successes with input provision and animal traction, together with recent high prices, are reflected in the relatively high weighted average returns to farmers in all three WCA countries. There is scope to reduce the high producer prices, allowing WCA cotton companies to break even, and still realize returns at least as high as those achieved by most ESA sectors.

The analysis also suggests that Zimbabwe is the best performer among ESA countries from a farmer's perspective, with Zambia and Tanzania hard to separate. However, given that the Tanzanian figures are based on a very good year, and that the Zambia labor input data are perhaps on the high side, therefore underestimating returns, Zambia probably has the edge. Thus, another key finding of this section is that, although competitive sectors within ESA have outperformed more coordinated ones on pricing (chapter 5), from a farmer's perspective they have not done so to such an extent as to outweigh their disadvantages in service provision.

Some doubts linger over input data for Mozambique (labor) and Uganda (hired services). However, apparently unattractive returns for the majority of farmers are consistent with disappointing medium-term production growth in both countries. A better-regulated local monopoly model should be able to deliver better services to more farmers than has so far been the case in Mozambique.

The question then becomes to what extent current performance at the farm level reflects the current state of the sector and how much it reflects the lagged impact of past sector performance. Our observation is that the lagged impact of past performance is large. This is clear both in WCA and in Zimbabwe, where Cottco's assistance through the early- to mid-2000s is still felt. WCA sectors continue to benefit from past investments, but have seen stagnant productivity growth since the mid-1980s (area planted has risen, but yield and net margins have stagnated or fallen).

It is, of course, difficult to predict future trends. The most confident prediction is that there will be limited change in the findings of this type of analysis over the next five years.

**Table 10.2 Summary of Average Yield and Return per Day of Labor**

| Country | Current sector type | Average yield (kg/ha) | Weighted average return per day of labor (US$) | Comments |
|---|---|---|---|---|
| Burkina Faso | Local monopoly | 1,088 | 1.54 | WCA sectors generate highest yields and returns. Returns still comparable to best in ESA when seed cotton price adjusted to "sustainable" level. |
| Cameroon | National monopoly | 1,167 | 1.34 | Return figure probably underestimated because of high labor input derived from SODECOTON data. |
| Mali | National monopoly | 1,030 | 1.70 | WCA sectors generate highest yields and returns. |
| Mozambique | Local monopoly | 575 | 0.48 | Return figure underestimated because of high labor input derived from focus groups, but low labor productivity and low wage rates are features of rural Mozambique. |
| Tanzania | Competitive | 556 | 1.09 | Competitive sector struggles to raise yields. Yields and returns both highly dependent on rainfall. |
| Uganda | Hybrid | 562 | 0.88 | Return figure probably underestimated because of high cost of hired services derived from focus groups. |
| Zambia | Concentrated | 671 | 0.91 | Return figure probably underestimated as a result of high labor input derived from focus groups. Average yield rising steadily since liberalization. |
| Zimbabwe | Concentrated | 910 | 1.23 | Best performer in ESA because of efforts over time to raise proportion of households in top producer groups. |

*Source:* Authors.

*Note:* These yield figures are the weighted average values derived from focus group discussions (group figures derived from SODECOTON monitoring data in the case of Cameroon).

# Annex A10  Methodology for Focus Group Discussion on Farmer Types

In several of the study countries, formal household surveys provide data on cotton yields and input use that can be stratified to show the performance of different categories of farmers. However, the difficulties of collecting reliable labor data in such surveys means that most of them do not contain information on labor use by different categories of cotton farmers.

To provide insights into labor use at modest cost, focus group discussions were undertaken in seven of the nine study countries (all except Cameroon and Benin). Discussions were undertaken in two (Burkina Faso, Mali) to six (Mozambique) villages per country, with efforts made to compare across regions or districts where there were considered to be important geographical differences in performance (Mozambique, Tanzania, Zimbabwe, Zambia) and also between more and less accessible villages within an area (Tanzania, Mozambique).

In all cases except Mozambique, a single group of informants provided information on cotton farmers in their village. There were commonly 5–10 informants per village. In Mozambique, a larger number of respondents were divided into groups (based on level of cotton production) with each group providing information on its own activities, albeit in the presence of people from other groups.

The first activity in each focus group discussion was a participatory ranking exercise based on the principles of wealth ranking. Where possible,[76] the name of every head of household in the village was recorded, then informants were asked to place the cards in piles based on the level of cotton production achieved by the household in a "normal" year. In most cases, this produced three or four piles. If one pile (typically the lower producers in ESA) was much larger than the others, the researchers asked for a further disaggregation of this group. In Mozambique, farmers were asked to divide themselves into groups and, after verification of each farmer's yield, each group responded in turn.

Once the groups had been identified, the informants were asked to describe the characteristics of households in each group, covering demography, income sources, and food production, as well as cotton production. This gave a picture of a typical household in each group. A crop budget was then drawn up for each group for one hectare (or acre) of cotton in a typical recent season.[77]

The overall assessment is that the focus group discussions were a cost-effective way of collecting reasonably reliable data on cotton production activities by different groups of farmers. However, the following issues are noted:

- The method tends to accentuate intergroup differences at the expense of intragroup variation. Thus, top groups are characterized as being able to rise above many of the problems that constrain poorer households and the impression is given that they uniformly achieve the best yields. Similarly, the dominant narrative for poorer households is that cotton yields and profitability are compromised by equipment, cash, or labor shortages.

- Current local issues may receive undue attention. One example was the legal requirement for Zimbabwean farmers to cut down their standing crops by August 15, following the harvest, which farmers in one area claimed was far too expensive to do because of labor shortages.

- If farmers are feeling dissatisfied with cotton (as many are for various reasons), they may inflate estimates of labor input to make a point. Facilitators need to have some grasp of comparative data from elsewhere to interrogate initial statements from focus group members where these statements appear questionable. With probing questions, plus comparisons across different groups within the village, informants will often arrive at a more considered final estimate.

- Where an extension agent is present during discussions, respondents may give the "correct"—but untrue—answers in line with extension recommendations. Questions on input use appear particularly susceptible to this distortion, as observed in Burkina Faso, Mali, and Zimbabwe. In one Zimbabwe focus group, researchers challenged the initial statements from group members about input use by the poorest groups. Eventually respondents admitted that the lower groups do not follow official recommendations and gave a "true" picture, which prompted the extension agent to confess to the researchers afterward that he had never realized this!

- Finally, there is a question of what constitutes a "normal" year. In Tanzania, 2003/04 and 2004/05 were bumper harvests, with national yields around 50 percent higher than the long-term average. Focus group estimates of normal performance reflected these recent good years. However, 2005/06 was a drought year, with national production a third of its level in the previous two seasons. As a result, seed cotton prices were driven up by companies scrambling to obtain scarce supplies. It is clear that an indicative budget should not combine the good yields of 2003/04 and 2004/05 with the high prices of 2005/06. In Mozambique a similar situation prevailed, with a near record harvest in 2005/06.

The issue of what constitutes a "normal" year can be handled by some initial data adjustments, combined with the construction of alternative scenarios for some countries. The data adjustments for the base cases are as follows:

- In Tanzania, reported yields were used (consistent with national performance in 2003/04 and 2004/05) in the budgets, but seed cotton prices were adjusted downward (from those recorded in 2005/06) to reflect plausible prices in a good harvest year. This is achieved by setting the average seed cotton price to 65 percent of the free-on-truck lint equivalent price (the share received by farmers in 2004/05). Seed cotton pricing in Tanzania works by companies agreeing on an opening price, based on a cautious estimate of the company's costs and desired profits, with market competition then driving the price up as the season progresses. The opening price is not heavily affected by the season, but when the harvest is good the price will be driven up less than in a drought season when companies are desperate for seed cotton to fulfill contracts. Thus, in adjusting prices to reflect a good season, the average price received by the top producers (who sell more of their product late in the season) is lowered by more than the average price received by the poorer producers (who are forced to sell much of their product as soon as the season opens to meet pressing cash needs).

- In Mozambique, budgets were constructed for four groups of farmers. A nationally representative annual household survey allows the distribution of cotton yields in both 2004/05 (a drought year) and 2005/06 (a good rainfall year) to be examined. The national yield spread is roughly mirrored by the following combination of focus group budgets: 2005/06—group 1 (40 percent), group 2 (20 percent), group 4 (40 percent); 2004/05—

group 1 (20 percent), group 3 (20 percent), group 4 (60 percent). Unlike the case of a competitive cotton sector such as Tanzania's, however, prices in Mozambique vary less between good and bad seasons. This book, therefore, considers just one scenario (figures 10.5 and 10.7), in which the balance of producers across groups is assumed to be as follows: group 1 (5 percent), group 2 (25 percent), group 3 (30 percent), group 4 (40 percent).

- In Mali, information on actual yields for the period 2004 to 2006 in the villages where focus group discussions were held was available from the local extension agent. The average village-level yield was similar to the average national yield (approximately 1,100 kg/ha), but higher than the weighted average yield generated by focus group discussions (approximately 1,000 kg/ha). For the budgets, therefore, the yields reported by the focus groups were adjusted upward by 10 percent.

## DISTRIBUTION OF PRODUCERS ACROSS GROUPS

Figure 10.5 shows estimates of the proportion of cotton farmers found within each group. These figures were determined in a number of ways. In Mali, Zambia, and Zimbabwe, the numbers were taken from the focus group exercises. In these countries, all farmers from each village were assigned to one group or another. The data are thus strictly representative of the focus group villages and at best only illustrative of the country more widely. Yet in each case, the weighted average yield figures that resulted from these proportions are plausible in light of other information for the country. In Burkina Faso and Cameroon, the data were from monitoring reports by SOFITEX and SODECOTON, respectively. In Tanzania, the numbers were taken from a 2004 cotton farmer survey (Maro and Poulton 2005), while those in Mozambique were based on a 2005 cotton farmer survey in two provinces spanning a wide range of typical yields in the country. In a number of countries, focus group respondents identified three groups of producers, while four groups were identified in others. Group 3 in Burkina Faso, Cameroon, Mali, and Zimbabwe correspond, therefore, to groups 3 and 4 in other countries.

A striking observation from figure 10.5 is that WCA sectors have a much higher proportion of households in groups 1 and 2 than ESA sectors. This reflects investments made over the years in the promotion of animal traction and use of fertilizer.

# Chapter 11: Cost Efficiency of Companies, Overall Sector Competitiveness, and Macro Impact

## *Nicolas Gergely*

This chapter brings together detailed data from ginning companies, along with farm-level data from the previous chapter, to calculate four additional outcome indicators: one for company performance (an intermediate outcome), one for overall sector competitiveness, and two indicators for the impact of the cotton sector on the countries' macro economies. The proposed typology generated clear predictions for only one of these indicators: ginning companies are expected to be the least efficient in monopolies, and the most efficient in competitive sectors. Other indicators depend on factors in which no single sector type is expected to consistently perform better.

The cost structures of cotton companies are based on accounting data disclosed by the companies in West and Central Africa (WCA) countries, and on interviews with ginners and other key informants in East and Southern Africa (ESA) countries.[78] To allow comparisons, costs do not include capital costs because these may be, depending on the ginner, financed either through equity or long- and medium-term loans. Cost estimates in most cases are from 2006, converted to US dollars at the prevailing exchange rate (calendar year 2005 for Mozambique). For Cameroon and Burkina Faso, however, costs correspond, respectively, to the 2004/05 and 2003/04 seasons because more recent data are not available (except producer prices, which are the 2006/07 actual prices). It has, however, been verified that costs have not dramatically changed in recent years in these countries, because domestic and imported inflation (mainly from imported equipment and oil) were more or less balanced by cost reductions. Benin is not included in this sample because recent reliable data were not available.

## Company Cost Efficiency

The indicator for company performance is the adjusted cost from farm gate prices to free-on-truck (FOT). This adjusted figure excludes taxes and the cost of critical functions, because these depend on policy and other factors unrelated to the efficiency of company operations. The value of seed is not deducted from the total cost figure because the performance of the seed market (and the value the companies can therefore get from seed sales) is also beyond the influence of the companies. These factors are brought back into the analysis when overall competitiveness and macro impacts of the cotton sector are considered.

This chapter focuses first on ginning costs as perhaps the key cost element in this indicator. Ginning costs in WCA (all national or local monopolies) range from US$0.134 to US$0.234 per kg of lint (table 11.1 and figure 11.1). In ESA, these costs are much lower—ranging from US$0.081 to US$0.123—for countries operating at reasonably high capacity utilization rates (Tanzania, Zambia, and Zimbabwe). For Mozambique and Uganda, which operate at about 20 percent of capacity, costs are comparable to those in WCA, at US$0.20 per kg and US$0.237 per kg, respectively. These are the only two countries in the region that do not allow open competition between ginners: Mozambique operates a local monopoly system, and Uganda operates a hybrid system with purchase quotas. Both these systems protect ginners from most competitive pressures and thus reduce

**Table 11.1 Comparative Analysis of Ginning Costs (US cents per kg of lint cotton)**

| Indicator | Burkina Faso[a] | Cameroon[c] | Mali[b] | Mozambique[d] | Tanzania[g] | Uganda[g] | Zambia[e] | Zimbabwe[f] |
|---|---|---|---|---|---|---|---|---|
| Type of system | Local monopoly | National monopoly | National monopoly | Local monopoly | Competitive | Hybrid | Concentrated | Concentrated |
| Exchange rate to US$ | 505 | 505 | 505 | 23.5 | 1,200 | 1,800 | 3,600 | Variable |
| Type of gins | saw | saw | saw | saw | roller | roller | saw | saw/roller |
| Average unit ginning capacity (tons) | 45,000 | 31,000 | 40,000 | 13,500 | 6,300 | 5,000 | 20,000 | 25,000 |
| % capacity utilized | 100 | 100 | 65 | 20 | 80 | 20 | 100 | 64 |
| **Fixed costs/kg of lint** | **5.84** | **4.03** | **8.00** | **17.15** | **1.84** | **12.29** | **4.69** | **3.29** |
| Depreciation | 3.31 | 3.06 | 4.59 | 7.81 | 0.65 | 6.02 | 2.50 | 1.90 |
| Salaries | 1.18 | 0.77 | 1.08 | 9.29 | 1.19 | 6.27 | 2.08 | 1.35 |
| Other | 1.35 | 0.20 | 2.32 | 0.05 | 0 | 0 | 0.11 | 0.05 |
| **Variable costs/kg of lint** | **9.99** | **9.39** | **15.39** | **6.51** | **6.31** | **7.66** | **7.61** | **4.76** |
| Energy | 2.50 | 3.07 | 4.40 | 2.36 | 0.94 | 3.04 | 0.50 | 0.04 |
| Packaging | 3.49 | 3.49 | 3.45 | 3.91 | 4.17 | 3.05 | 3.50 | 2.17 |
| Other (including maintenance) | 4.00 | 2.84 | 7.54 | 0.24 | 1.20 | 1.58 | 3.61 | 2.56 |
| **Total cost/kg of lint** | **15.83** | **13.42** | **23.39** | **23.66** | **8.15** | **19.96** | **12.30** | **8.06** |
| At 100% capacity | 15.83 | 13.42 | 20.59 | 9.94 | 7.78 | 10.12 | 12.30 | 6.88 |
| At mean 1995–2006 exchange rate [h] | 13.62 | 11.55 | 17.71 | n.a. | n.a. | n.a. | n.a. | n.a. |

*Source:* Authors.

*Notes:* n.a. = Not applicable because no need to adjust for different exchange rate.
   Both Zambia and Uganda use some secondhand ginning equipment.
a. SOFITEX actual accounts for 2003/04.
b. CMDT budget for 2006/07.
c. SODECOTON actual account for 2004/05.
d. Estimate for 2005 calendar year (Boughton 2008).
e. Estimates by Gerald Estur (consultant) for 2005/06 (ginners contend they are underestimated).
f. Estimates for 2005/06 (Poulton and Hanyani-Mlambo 2007).
g. Estimates based on 2006/07 costs but 2004/05 capacity utilization (Poulton and Maro 2007).
h. CFA franc 587/US$; costs at actual capacity.

incentives for cost containment. At 100 percent capacity utilization rates in all countries, there would be no overlap in ginning costs between the two regions: WCA countries would range from US$0.13 to US$0.20 per kg, while those in ESA would range from US$0.07 to US$0.12 per kg.

A combination of technical and structural factors likely contributes to this stark difference in ginning costs between the two regions. First, WCA uses only saw gins. Investment costs (and hence depreciation costs) for saw gins are substantially higher than for roller gins. Roller gins are predominant in Uganda and also widely used in Tanzania.[79] In addition, a number of ginners in Tanzania and Uganda import equipment from India, often secondhand. Some Zambian ginners install used saw gins. All this equipment is much less expensive than the US or European equipment purchased by WCA cotton companies. Second, energy is much cheaper in Zimbabwe (less than US$0.001 per kg of lint),[80] Zambia (US$0.005 per kg), and Tanzania (US$0.009 per

**Figure 11.1 Estimated Average Ginning Costs at 2006 Capacity Utilization Rates in Study Countries**

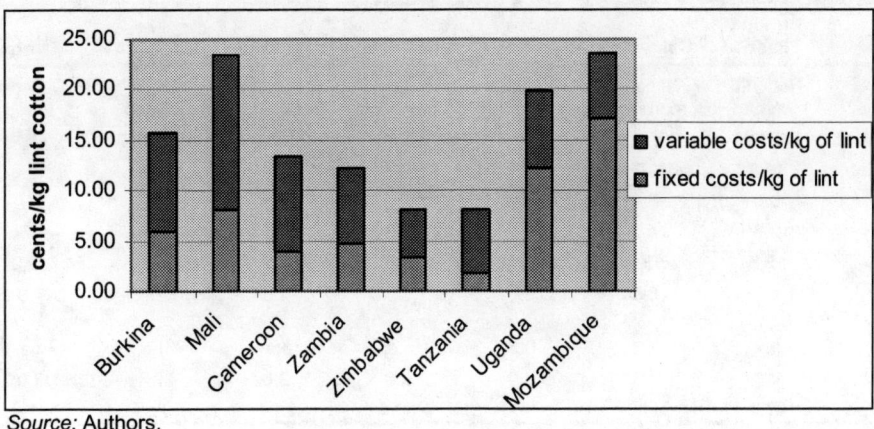

*Source:* Authors.

kg), than in WCA countries (around US$0.025 per kg). Finally, maintenance and other variable costs are, as a general rule, substantially lower in ESA, reflecting higher cost efficiency at the processing stage.

The gap between ginning costs in WCA and ESA countries would be reduced, but not eliminated, if the CFA franc/US$ exchange rate returned to its 1995–2006 mean, which corresponds to the period since the 1994 devaluation (see final line of table 11.1): costs in Cameroon would be at the upper end of those in ESA, but costs in Burkina Faso, and especially Mali, would remain much higher. This suggests that, while the current CFA franc/US$ exchange rate does contribute to the current financial crisis in the region, a more fundamental cause for the financial crisis, at least in Burkina Faso and Mali, is very high costs of operation in cotton ginning. This conclusion becomes even stronger when one realizes that figures for the highest cost ginner in ESA—Zambia—are heavily affected by the sharp appreciation of the kwacha during 2005/06. Zambia's ginning cost in US dollars would be below US$0.10/kg of lint if the kwacha were at its 2005 level.

Table 11.2 builds in additional costs from farm gate to FOT to develop the overall indicator of company cost efficiency: total farm gate to FOT costs excluding taxes and critical functions. The main discriminating factor for collection costs appears to be the size of ginning units; transport costs are lower in countries with smaller ginning units (particularly Uganda) because the purchasing area for each unit is smaller, and higher in countries with large-scale ginning units (Zambia and WCA).

Short-term financial costs are higher in monopolistic systems (except Cameroon) for managerial reasons. In such systems, cotton companies tend to buy seed cotton immediately after harvest, and to sell throughout the year, thus holding high average stocks. Ginners in competitive systems tend to minimize their stocks by selling immediately after processing. Cameroon is unique for monopolistic systems—most of its stock, at least through the end of the 2006 season, was financed by cash reserves of the company and the farmer organization.

**Table 11.2 Company Performance Indicator: Adjusted Costs, Farm Gate to FOT**
*(US¢ per kg of lint)*

| Indicator | Burkina Faso | Cameroon | Mali | Mozambique | Tanzania | Uganda | Zambia | Zimbabwe |
|---|---|---|---|---|---|---|---|---|
| Type of cotton industry | National monopoly | National monopoly | National monopoly | Local monopoly | Competitive | Hybrid | Concen-trated | Concen-trated |
| Date of data | 2003/04 | 2004/05 | 2006/07 | 2005 | 2006/07 | 2006/07 | 2005/06 | 2005/06 |
| Collection of seed cotton | 10.3 | 9.7 | 7.3 | 9.7 | 9.4 | 7.9 | 13.6 | 8.1 |
| Transport | 6.5 | 5.0 | 5.2 | 8.8 | 5.1 | 3.2 | 7.7 | 6.1 |
| Other | 3.8 | 4.7 | 2.1 | 0.9 | 4.1 | 4.8 | 5.9 | 2.0 |
| Ginning costs | 15.8 | 13.4 | 23.4 | 23.7 | 8.2 | 20 | 12.3 | 8.1 |
| Overhead | 4.6 | 6.5 | 6.0 | 3.0 | 1.8 | 2.5 | 4.0 | 4.9 |
| Portion of overhead paid to DAGRIS | 1.2 | 1.2 | 0.3 | n.a. | n.a. | n.a. | n.a. | n.a. |
| Short-term financing cost | 6.7 | 1.3 | 8.0 | 2.0 | 2.0 | 1.6 | 2.0 | 3.6 |
| Total adjusted costs | 37.4 | 30.8 | 44.6 | 38.4 | 21.4 | 31.9 | 31.9 | 25.5 |

*Source:* Authors.

*Note:* n.a. = Not applicable.

a. Cost of input borne by cotton company.

Overhead costs are consistently higher in the WCA monopoly systems, where cotton companies are larger, combine a broader scope of functions, and lack incentives to minimize costs. Overhead costs are lower in competitive and concentrated systems, where companies have more incentives to minimize costs and also focus on a narrower set of functions.

Focusing now on the final indicator of company performance—total adjusted costs (final line of table 11.2, and figure 11.2)—companies in concentrated and competitive systems show clear evidence of greater efficiency, especially in the more competitive sectors. Burkina Faso, Mali, and Mozambique, all monopoly sectors, show the highest adjusted farm gate to FOT costs, with costs in Mali being especially high. Costs in Tanzania's competitive sector and Zimbabwe's increasingly competitive sector are substantially below all other countries, although the Zimbabwe figure is heavily influenced by the dramatic depreciation in the real exchange rate since the onset of economic crisis in 2001. Based on the high share of roller gins in Uganda, its costs should be at least as low as those in Tanzania, but instead Uganda's costs are comparable to those in Zambia (which uses saw gins) and Cameroon. Very low capacity utilization, perpetuated by the hybrid regulatory structure, drives this result. Zambia's figure, which is the highest of the market-based systems, includes costs involved in achieving a very high quality premium. Cameroon shows what a national monopoly can achieve when well managed and left relatively free from political influence, yet its costs are still well above those in Zimbabwe and Tanzania.

## Overall Competitiveness

The overall competitiveness indicator is the ratio of total FOT costs to total FOT value in each sector. The cost side of this indicator is developed by starting with the adjusted farm gate to FOT costs from table 11.2, and adding purchase price; profit taxes; payments for critical functions such as extension; input subsidies paid by companies (if any); and anything paid by the companies for research, road maintenance, and other public goods. Any taxes included in these costs are not deducted. Company revenues can only be theoretically estimated because information is not publicly available on actual selling prices and costs except in WCA's national

**Figure 11.2 Company Performance Indicator: Adjusted Total Cost, Farm Gate to FOT, 2006/07**

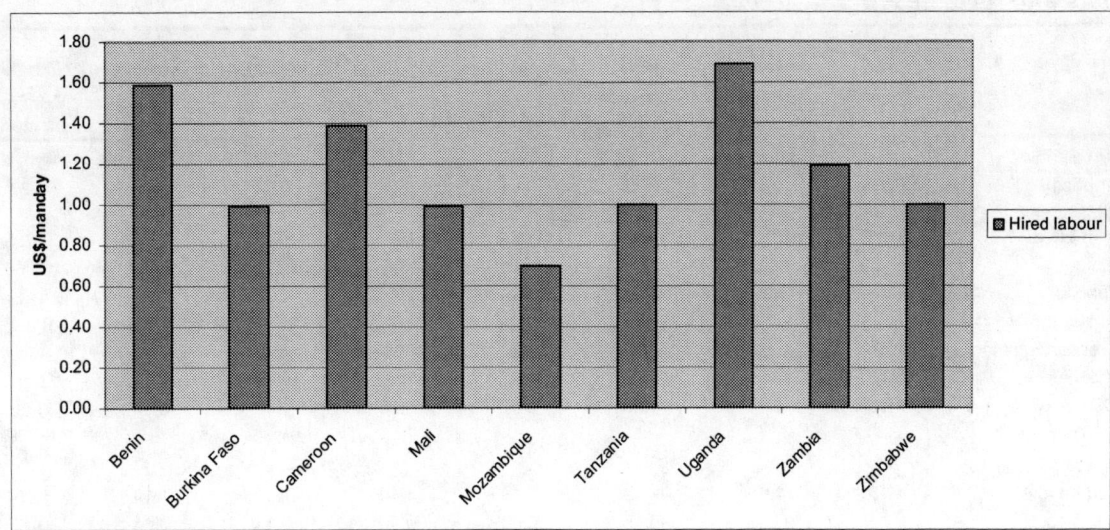

*Source:* Authors.

monopolies. The calculations are based on Cotlook A Index prices, quality premiums as estimated in chapter 7, and information on the value of seed sales from interviews in each country.

Results are shown in table 11.3. Ratios above 1.0 indicate that the sector is generating deficits, unable to cover costs at estimated realized export prices and sales prices of seed. The WCA monopolies look very poor, with even Cameroon less competitive than any of the ESA countries. High producer prices and high costs for critical functions (lower only than those in Uganda), combine with the high operating costs already documented in table 11.2 to drive this result. Remarkably, Mozambique now looks very competitive, but this is in large measure due to the very low prices that this poorly regulated local monopoly sector pays to its farmers. Zambia emerges as perhaps the most competitive sector, in large measure as a result of the very high price premium it now earns on international markets. Zimbabwe's revenues are reduced by the requirement that 30 percent of output be sold on the local market at lower prices; yet Zimbabwe's sector still emerges as relatively competitive.

In seven of the eight countries in this analysis, cotton companies pay for all or a significant portion of extension. The one exception is Tanzania, which is the only competitive sector; companies spending money on extension would be at a competitive disadvantage in such a system and so do not provide this service. Uganda has a competitive structure, but its regulatory framework (which classifies its sector as a hybrid) allows companies to cooperate in supporting extension (supplemented by donor money). These company costs are certainly recovered from farmers through the buying price, as they must be in other ESA countries. The sector deficits in the WCA region suggest that these (and other) costs are not being fully recovered. Extension costs are higher in WCA than in ESA (US$0.025 to US$0.040 per kg of lint, compared with zero to US$.029 per kg of lint in ESA), and are highest in Mali, because the scope of extension includes rural development functions beyond cotton. Extension in Cameroon also covers rural development, but its cost is shared between the cotton company and the farmer association. The

**Table 11.3 Overall Competitiveness Indicator: FOT Costs/FOT Revenue, 2006/07**
*(all costs and revenues in US$/kg lint)*

| | Burkina Faso | Cameroon | Mali | Mozambique | Tanzania | Uganda | Zambia | Zimbabwe |
|---|---|---|---|---|---|---|---|---|
| | Local monopoly | National monopoly | National monopoly | Local monopoly | Competitive | Hybrid | Concen-trated | Concen-trated |
| **Costs** | | | | | | | | |
| Producer price/kg lint | 0.78 | 0.78 | 0.75 | 0.51 | 0.69 | 0.71 | 0.64 | 0.67 |
| Adjusted farm gate to FOT costs (from table 11.2) | 0.374 | 0.308 | 0.446 | 0.384 | 0.214 | 0.319 | 0.319 | 0.255 |
| Direct taxes | n.a. | 0.004 | n.a. | 0.022 | 0.037 | 0.026 | 0.012 | 0.006 |
| Critical functions | | | | | | | | |
| Extension | 0.025 | 0.027 | 0.040 | 0.007 | n.a. | 0.027 | 0.015 | 0.029 |
| Roads, research, other | 0.004 | 0.015 | 0.010 | n.a. | 0.019 | n.a. | n.a. | n.a. |
| Input subsidies | n.a. | n.a. | 0.018 | n.a. | n.a. | 0.084 | n.a. | n.a. |
| Total FOT costs | 1.18 | 1.13 | 1.26 | 0.92 | 0.96 | 1.17 | 0.99 | 0.96 |
| **Income** | | | | | | | | |
| FOB value (CFR[b] minus freight and handling) | 1.18 | 1.18 | 1.18 | 1.18 | 1.19 | 1.19 | 1.19 | 1.05[a] |
| FOT to FOB costs | 0.15 | 0.19 | 0.15 | 0.03 | 0.10 | 0.13 | 0.14 | 0.11 |
| FOT value | 1.03 | 0.99 | 1.03 | 1.15 | 1.09 | 1.06 | 1.05 | 0.94 |
| Quality premium | 0.04 | 0.07 | 0.02 | -0.04 | -0.01 | 0.04 | 0.11 | 0.06 |
| Sale of seed/kg lint | 0.06 | 0.09 | 0.05 | 0.05 | 0.08 | 0.16 | 0.11 | 0.13 |
| Total Income, FOT | 1.13 | 1.15 | 1.10 | 1.16 | 1.16 | 1.26 | 1.27 | 1.13 |
| **FOT cost/FOT revenue** | 1.05 | 0.99 | 1.15 | 0.80 | 0.83 | 0.93 | 0.78 | 0.85 |

*Source:* Authors.

*Note:* n.a. = Not applicable.

a. Taking into consideration sales on domestic market (30 percent of sales) at a lower administered price.

b. CFR = Costs, insurance, freight.

monopoly cotton companies also contribute to road maintenance and research costs, though probably less in Mozambique than in the WCA countries.[81]

## Macro Impacts

The two macro impact indicators are per capita total value added from the cotton sector and per capita net budgetary contributions of the sector. Value added at the farm level is calculated by subtracting nonlabor production costs from the farm gate value of seed cotton production. In the absence of mean figures across all farmers, the data for the farmer group that contains the median farmer are used (see table 11.4 and figure 11.3).

Value added is calculated using two different definitions of nonlabor production costs. In the first case, input costs are deducted from the gross value of seed cotton production. In the second case, costs of animal traction and motorized services are also deducted, although the latter are extremely rare. The most appropriate definition can be debated. The argument for deducting animal traction costs from value added is that they include rental or amortization of equipment, plus veterinary and feeding costs for oxen. These intermediate inputs account for about 80 percent of the estimated animal traction costs in WCA sectors. The argument for not deducting animal traction costs is that oxen (and, indeed, equipment) are assets that farm households have accumulated in large part through their engagement with cotton production. Deducting animal traction costs would have the perverse effect of lowering the estimates of the value added generated by WCA sectors (because these sectors have proceeded further with animal traction than ESA) when this progress owes much to cotton.

Value added at the ginnery stage can be estimated by subtracting all nonlabor and nontax costs for purchase and collection of seed cotton and for processing and marketing of lint cotton from the total FOT value of lint and seeds. The two value-added numbers are then summed and the result is converted to per capita figures by multiplying by production and dividing by total country population (table 11.4, figure 11.3).

WCA countries perform relatively poorly on total value added per kg of lint. Yet Burkina Faso and Mali perform quite well when this number is adjusted to the country's total population; the two different treatments of animal traction costs have no impact on overall conclusions. The high value added per capita in Burkina Faso and, to a lesser degree Mali, flows directly from the orientation of the national monopolies (now local monopolies in Burkina Faso) to extend cotton cultivation to all farmers in areas deemed adapted to cotton cultivation. In ESA, Zambia and Zimbabwe perform very well because they achieve high value added per kg and also reach a relatively high share of the population (though much less than in WCA). Tanzania's sector, in contrast, reaches far fewer farmers as a share of population, and its value added per capita thus

**Table 11.4 Total Value Added per Capita from Cotton Sector, 2006**

| Indicator | Burkina Faso | Mali | Cameroon | Mozambique | Zambia | Zimbabwe | Tanzania | Uganda |
|---|---|---|---|---|---|---|---|---|
| **Value added at farm level** | | | | | | | | |
| Seed cotton yield (kg/ha) | 1,100 | 1,011 | 1,120 | 565 | 600 | 800 | 438 | 563 |
| Seed cotton price (US$/kg) | 0.33 | 0.32 | 0.32 | 0.21 | 0.25 | 0.29 | 0.24 | 0.25 |
| Gross value per ha (US$) | 359 | 324 | 358 | 119 | 150 | 232 | 105 | 141 |
| Deduction for input costs (US$/ha) | 165 | 160 | 133 | 14 | 20 | 90 | 6 | 8 |
| Deduction for hired services (US$/ha) | 35 | 31 | 16 | 1 | 10 | 25 | 1 | 71 |
| Value added/ha, I (US$) | 195 | 164 | 225 | 105 | 130 | 142 | 99 | 133 |
| Value added/ha, II (US$) | 160 | 133 | 209 | 104 | 120 | 117 | 98 | 62 |
| Gross outturn ratio (percent) | 42 | 42 | 41 | 39 | 40 | 41 | 36 | 35 |
| Value added/kg lint, I (US$) | 0.42 | 0.39 | 0.49 | 0.47 | 0.54 | 0.43 | 0.63 | 0.67 |
| Value added/kg lint, II (US$) | 0.35 | 0.31 | 0.46 | 0.47 | 0.50 | 0.36 | 0.62 | 0.31 |
| **Value added at ginning level (US$/kg lint)** | | | | | | | | |
| Total FOT income (table 11.3) | 1.13 | 1.10 | 1.15 | 1.16 | 1.27 | 1.13 | 1.16 | 1.26 |
| Deduction for price paid to farmer/kg lint | 0.78 | 0.75 | 0.78 | 0.51 | 0.64 | 0.67 | 0.69 | 0.71 |
| Deduction for other nontax, nonlabor costs | 0.30 | 0.39 | 0.23 | 0.29 | 0.22 | 0.22 | 0.16 | 0.24 |
| Value added/kg lint | 0.05 | −0.04 | 0.14 | 0.36 | 0.41 | 0.24 | 0.31 | 0.31 |
| **Total value added** | | | | | | | | |
| Total value added/kg lint, I (US$/kg lint) | 0.47 | 0.35 | 0.63 | 0.83 | 0.95 | 0.67 | 0.94 | 0.98 |
| Total value added/kg lint, II (US$/kg lint) | 0.40 | 0.27 | 0.60 | 0.83 | 0.91 | 0.60 | 0.93 | 0.62 |
| Production, thousand metric tons lint | 311 | 186 | 89 | 43 | 80 | 123 | 126 | 22 |
| Total value added I (thousand US$) | 146,513 | 64,188 | 56,146 | 35,902 | 76,133 | 82,770 | 118,265 | 21,641 |
| Total value added II (thousand US$) | 123,222 | 50,609 | 53,045 | 35,707 | 72,800 | 73,395 | 117,466 | 13,714 |
| Population, 2005 (thousand) | 13,933 | 11,611 | 17,795 | 20,533 | 11,478 | 13,120 | 38,478 | 28,947 |
| Total value added per capita, I (US$) | 10.52 | 5.53 | 3.16 | 1.75 | 6.63 | 6.31 | 3.07 | 0.75 |
| Total value added per capita, II (US$) | 8.84 | 4.36 | 2.98 | 1.74 | 6.34 | 5.59 | 3.05 | 0.47 |

*Source:* Authors.

*Note:* Value added I = Gross value of seed cotton production minus input costs. Value added II = Value added I minus costs of animal traction and motorized services.

**Figure 11.3 Total Value Added per capita by Cotton Sector, 2006/07**

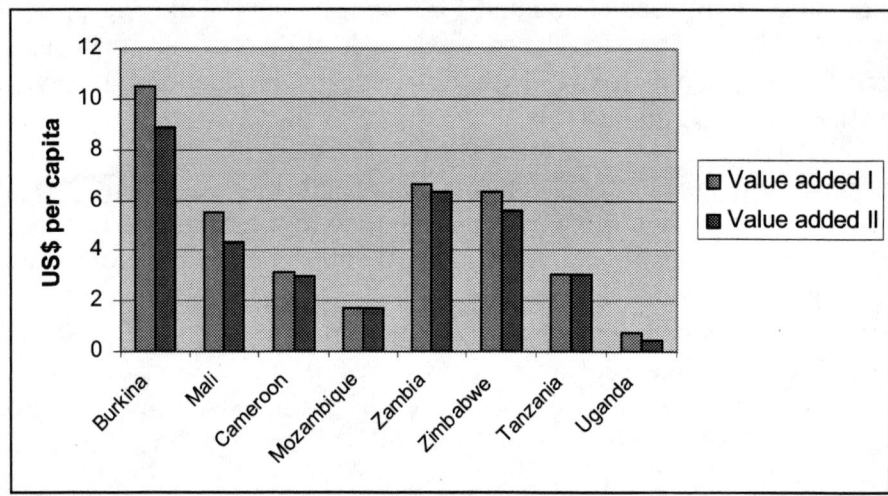

*Source:* Authors.

*Note:* Value added I = Gross value of seed cotton production minus input costs. Value added II = Value added I minus costs of animal traction and motorized services.

falls to about half that in Zambia and Zimbabwe. Mozambique and Uganda perform the poorest, because of chronically low production in each country.

The final indicator shows that in Burkina Faso and Mali, these positive results for value added currently come at a very high cost to the national budget (table 11.5). Mali's negative budgetary contribution is especially striking when compared with the value-added figures in table 11.3: for every dollar of value added generated, the Malian government had to pay roughly US$0.50 from its budget in 2006. Burkina Faso's figure may be less alarming in total and when compared with value-added figures, but caused major problems for the country and its ginning companies in 2006 and 2007. Within ESA, the concentrated and competitive sectors (Tanzania, Zambia, and Zimbabwe) clearly make the greatest contribution to the state budget.

Over all three macro indicators, the concentrated and competitive systems in Tanzania, Zambia, and Zimbabwe emerge as the strongest performers: company costs are among the lowest, overall competitiveness is strong (without punishing farmers with very low prices, as in Mozambique), they generate substantial value added per capita, and they do all of this while making by far the highest contributions to the national budgets.

**Table 11.5 Net per Capita Budgetary Contribution of Cotton Sector, 2006**

|  | Burkina Faso | Mali | Cameroon | Mozambique | Zambia | Zimbabwe | Tanzania | Uganda |
|---|---|---|---|---|---|---|---|---|
| Total direct + indirect taxes paid (thousand US$) | 4,261 | 2,883 | 1,388 | 1,471 | 1,904 | 2,312 | 5,531 | 719 |
| Budgetary transfers received (thousand US$) | −15,550 | −31,620 | 0 | 0 | 0 | 0 | 0 | 0 |
| Net budgetary contribution | −11,289 | −28,737 | 1,388 | 1,471 | 1,904 | 2,312 | 5,531 | 719 |
| Population, 2005 (thousand) | 13,933 | 11,611 | 17,795 | 20,533 | 11,478 | 13,120 | 38,478 | 28,947 |
| Net budgetary contribution, per capita | −0.81 | −2.47 | 0.08 | 0.07 | 0.17 | 0.18 | 0.14 | 0.02 |

*Source:* Authors.

*Note:* Budgetary transfers assumed equal to losses of public companies—CMDT in Mali, SOFITEX in Burkina.

# Chapter 12: Conclusions
## *Patrick Labaste, Colin Poulton, and David Tschirley*

Implicit in reform programs in cotton and other sectors in Africa to date has been the notion that structure matters, at least insofar as it promotes or impedes competition. However, relatively little sustained attention has been paid to the precise structure that post-reform sectors could take or why a particular structure might be preferable.[82] This present work strongly confirms that sector structure matters to performance and to the types of approaches needed to improve performance. At the same time, it recognizes that other factors, such as history, managerial competence, geography, politics, macroeconomics, developments in competing or complementary sectors, and even luck, also play roles that cannot be ignored.

Despite the financial crisis currently facing a number of cotton sectors in WCA, experts tend to agree that the high intrinsic quality of its fiber, the fact that it is handpicked, and the low unit production costs of its smallholder production base still give African cotton important growth potential on the world market in the long run. However, remaining competitive and maintaining or increasing market share will require considerable effort geared toward narrowing existing gaps (in productivity, quality management, and so forth) and building more on comparative advantages. These objectives are important notwithstanding factors beyond the direct control of Sub-Saharan African (SSA) governments and stakeholders, such as the evolution of the euro/US$ exchange rate and market distortions from subsidies in OECD countries and in China.

This final chapter summarizes the study's main findings on the links between sector types and performance outcomes. It then suggests ways forward for African cotton sectors, looking first at themes that cut across sector types, then individually at each sector type.

## Summary of Expected and Realized Performance across Key Indicators

Table 12.1 brings together information on the selected performance indicators by sector type, comparing expected with realized performance. The expectations generated from the typology in chapter 4 about the strengths and weaknesses of different sector types are borne out in a number of ways, if not in all cases.

With regard to sector structure, four broad conclusions emerge from the study:

First, no one sector type performed sufficiently well on all performance indicators and throughout the last 10–15 years to be considered unambiguously "best." The study has revealed strengths and weaknesses in all systems, particularly when they are examined over an extended period. All cotton sectors are seeking to adapt by implementing reforms.

Second, while no single benchmark model emerges from the countries under review, useful distinctions can and must be made. Current weaknesses of some of the West and Central Africa (WCA) sectors—lack of financial sustainability, low economic efficiency, and stagnating productivity at farm level—threaten their short-term survival. At the other end of the spectrum, competitive models like Tanzania and Uganda are apparently stuck with low productivity and (in unregulated Tanzania) low quality, plus long-term concerns about resource depletion and an inability to make a significant dent in poverty rates. By contrast, the concentrated and, less clearly, the local monopoly systems seem to have the greatest capacity to tackle their current difficulties, develop regulatory systems, and deliver improved performance in the near future. Sector type does matter.

Third, while there has been considerable fluidity in the structure of African cotton sectors over the past 10–15 years (see figure 4.2), there is also an element of path dependency in sector structure.[83] It is difficult to consolidate a sector from 30 or more ginning companies to five or six. Conversely, it is not advisable and realistic for a country to try to change overnight from a highly coordinated national monopoly, with its beliefs and operating procedures, to a fragmented, competitive model. Where needed, systemic reforms should, therefore, be designed to shift sector structure progressively toward a preferred alternative model, rather than radically altering the structure over a short period; where systemic reform is not called for, the priority is to seek new institutional arrangements that make the existing system work better. Both of these issues are examined in more concrete terms later in this chapter.

Fourth, while some African cotton sectors may be faring better than others, on a global scale all are lagging behind the best performers in the world. Global competitiveness must therefore be kept high on the agenda in every African cotton sector. Though a number of SSA producers have succeeded in expanding their share of the world lint market since the mid-1990s, they face increasing competition from other countries and from synthetic fibers, hence they should expect low future prices. This means that they must continually cut costs, raise productivity, and add value if they are to maintain attractive returns to producers and to make a positive contribution to national poverty reduction goals. Whether this needs to be done through systemic reform or through institutional innovation within the existing structure depends on the country's specific circumstances. In either case, countries need to address some common factors hindering competitiveness by improving research and extension responsiveness and efficiency, technology transfer in areas such as dissemination of genetically modified varieties, and natural resource conservation.

### PERFORMANCE ON CORE ACTIVITIES AND SERVICE DELIVERY: THE PROCESS INDICATORS

The typology presented in this book generated clear expectations regarding four of the six selected process indicators (quality, pricing, input provision, and extension), and in all four cases, expectations were largely confirmed (table 12.1).

Concentrated sectors perform best on quality. They also provide input credit and extension advice to large shares of cotton farmers, although farmer coverage is not as complete as in national and local monopolies. Credit repayment rates in concentrated systems are typically high, but less stable than in national monopolies. The 1990s showed that concentrated systems can deliver attractive prices to farmers. However, post-2000 performance has been poor, as might be feared in oligopolies.

National and local monopolies in WCA countries provide input credit and extension to all farmers and achieve high and fairly stable credit repayment rates. However, the quality of extension assistance in these monopoly systems has probably declined since the mid-1980s.

116

**Table 12.1 Summary of Expected and Realized Performance across Key Indicators**

| Type of indicator | Indicator | Expected performance | Realized performance | Comments |
|---|---|---|---|---|
| **Process indicators** | | | | |
| Quality | Estimated average realized premium over Cotlook A Index on world markets (US$/lb lint) | Best in concentrated sectors. Lowest in competitive. Conditions in national and local monopolies should be conducive to quality, but performance depends on management culture and regulatory effectiveness. | Expectations largely confirmed in ESA market-based sectors; Zambia (most concentrated) + 0.05/lb, Zimbabwe + 0.03/lb, Uganda + 0.02/lb, Tanzania (most competitive) - 0.01/lb. State monopolies in WCA lower than concentrated sectors in ESA, at par to + 0.03/lb; Mozambique (local monopoly) worst at - 0.02/lb. | Mozambique performance is highly variable across companies. Ineffective regulation and previous dominance of original investors (with little focus on productivity and quality) still affecting overall premiums.<br><br>Zimbabwe performance has fallen with new entrants.<br><br>Mali is the poorest performer in WCA because of management issues. |
| Pricing | Mean percentage of free-on-truck paid to farmers | Best in competitive systems. Lower in concentrated, market-based sectors. In national and local monopolies, depends on political factors and bargaining strength of farmer organizations. | Expectations largely confirmed outside monopoly sectors: Tanzania (most competitive) 70%, Uganda 68%, Zimbabwe (more concentrated) 49%, Zambia (most concentrated) 55%. National monopoly systems all paid 71%—76% in 2000–05, but much lower levels in 1990s. Mozambique, with least powerful farmer groups and local monopoly system, is lowest at 48%. | High share of world price in competitive systems (Tanzania, Uganda).<br><br>High prices to farmers in WCA are not sustainable.<br><br>Prices exceptionally low in Zimbabwe in early 2000s, but impact of new entry on price still to be proven. |
| Input provision | • Percentage of cotton farmers receiving input credit<br><br>• Adequacy and quality of input credit package, if provided<br><br>• Repayment rate | National and local monopolies best on provision and repayment, indeterminate on adequacy and quality. Concentrated, market-based sectors next best on provision, possibly with better incentives on adequacy and quality. Competitive sectors unable to provide input on credit. | Expectations largely confirmed:<br>• % receiving input credit: ≈ 100% in all WCA countries, Mozambique, and Zambia; 90%–95% in Zimbabwe (2006/07); 0% in Tanzania. Subsidized cash sale in Uganda (where past experiments with credit unsuccessful).<br>• Adequacy worst (though variable) in Mozambique; best in WCA, Zimbabwe (before 2003), and Uganda (under hybrid system)<br>• Credit repayment consistently >90% in WCA; highly variable in Mozambique, Zambia; 90% but falling in Zimbabwe. | WCA and (to a lesser extent) Zimbabwe, Uganda provide fertilizer; Uganda's provision enabled by hybrid system imposed on competitive structure.<br><br>Adequacy falling in Zimbabwe, very low among new entrants. Tanzania able to provide small amounts of input through forced savings mechanism. |

*(continued)*

117

| Type of indicator | Indicator | Expected performance | Realized performance | Comments |
|---|---|---|---|---|
| Extension | • Provision of assistance<br>• Qualitative assessment of quality | Same as input provision, though extension in general expected to be weaker than input credit provision. | Expectations largely confirmed, with exception of Mozambique, where some companies provide almost no extension | Measurement very difficult, and distinction between providing input and extension advice is not always clear. |
| Research | • Number of varieties released and taken-up in past 10 years<br>• Use of new technologies (GM varieties, low-volume herbicides) | No clear prediction, though monopoly and concentrated systems may have better company input into breeding work | No clear pattern. Mali (6) and Zimbabwe (4) have most releases. WCA has seen much greater organizational stability in research in the past, because of persistence of national monopoly approach. | Nature of agricultural research means that history exerts a big influence. Impact of sector type limited by fact that state still controls cotton research in most countries. |
| Valorization of by-products | Value of cotton seeds | No clear prediction—only in WCA does history of vertical integration establish a systematic link with ginning structure. | Abnormally low in Mali and Burkina Faso (monopsony in cottonseed oil markets). High prices realized in other landlocked countries. | Both demand (for oil and cake) and policy factors beyond the control of ginners exert a big influence. |
| **Intermediate outcome indicators** | | | | |
| Farm-level performance | • Mean yield past 5 years (kg seed cotton/ha)<br>• Trend, past 10 years | Expected performance on yield level strongly related to expectations for input provision and extension. No clear prediction for yield trend. | Mean yields remain highest in WCA; within ESA, concentrated sectors perform best. However, yield stagnation in WCA since mid-1980s. Zambia (concentrated) provides best example of yield growth in ESA, but overall ESA yields have failed to close gap on world average. | Current yields heavily influenced by past investments, especially in research and extension, so typology alone does not explain performance variations. |
| Company performance | Adjusted farm gate to FOT cost (US$/kg lint) | Worst in monopolies, best in competitive sectors. | Expectations largely confirmed: US$0.31–0.45 in WCA monopolies, US$0.38 in Mozambique; in market sectors, US$0.21 in Tanzania to US$0.32 in Zambia and Uganda. | Hybrid regulatory structure in Uganda protects inefficient companies. Roller gins and competition in Tanzania keep costs down. |
| **Overall outcome indicators** | | | | |
| Farmer welfare | Returns per day to family labor (US$/day) | No clear prediction | Generally highest in WCA national and local monopolies, US$1.30–US$1.70<br>Highly variable in ESA, US$0.79 in Tanzania (competitive) to US$2.11 in Zimbabwe (concentrated).<br>Worst in Mozambique (local monopoly) at $0.44. | Past investments in animal traction important. Current prices not sustainable in WCA, but yields a bigger determinant of returns than prices. |

(continued)

118

| Type of indicator | Indicator | Expected performance | Realized performance | Comments |
|---|---|---|---|---|
| Overall competitiveness | Ratio of total FOT cost to total FOT value | No clear prediction. | WCA monopolies least competitive, 0.98 (Cameroon) to 1.15 (Mali). ESA concentrated and competitive much more competitive at 0.76 (Zambia) to 0.88 (Zimbabwe). 0.79 in Mozambique. | Low (that is, competitive) figure in Mozambique, partly result of very low prices to farmers. Low figure in Zambia partly result of not passing quality premium on to farmers. WCA has high farm prices and high ginning costs. |
| Macro impact | • Total value added per capita<br>• Net budgetary contribution per capita (taxes paid minus transfers received) | Traditionally thought to be best in WCA monopolies, where coverage of farmers greatest and yields highest. | *Total value added.* No clear pattern by sector type, though concentrated market-based systems in ESA clearly best in that region, with Zambia and Zimbabwe ≈ US$6 per capita; Burkina Faso highest of all at US$8.84. per capita; Uganda and Mozambique lowest because of low production. *Net budgetary contribution.* Very poor in non-market-based systems: Mali government paid net US$2.47 per capita to sector in 2006, Burkina paid US$0.81; sectors in Cameroon, Mozambique, and Uganda made positive but small contributions (US$0.02–US$0.08); market-based sectors (Tanzania, Zimbabwe, Zambia) all made positive contributions of US$0.14–US$0.18 per capita. | *Value added.* Tanzania (competitive) has highest value added per kg lint, but highly variable production means macro contribution will also vary greatly. Zambia and Zimbabwe (concentrated) show much less variation. |

*Source:* Authors.

*Note:* FOT = free-on-truck; GM = Genetically modified.

While seed cotton prices were at their lowest in the 1990s, since 2000, they have benefited from new price-setting mechanisms, including greater involvement of farmer representatives in price negotiations. However, they are now at unsustainably high levels. Performance on lint quality is variable. In general, these monopoly systems underperform on quality management, in part because of political influence within the cotton sector.

Meanwhile, Mozambique's local monopoly system underperforms expectations more generally, although there is considerable divergence in performance across concession zones, with newer entrants often outperforming early incumbents. Mozambique has the highest quality discount on world markets and, in several concession areas, the poorest input supply and extension assistance outside of Tanzania. This performance is less surprising when one considers that the country emerged from a devastating civil war only in the early 1990s, and that its regulatory capacity remains very low.

Evidence is strong in Tanzania and Uganda that, within market-based systems, competition increases prices paid to farmers. Despite the recent high prices paid by the WCA monopoly systems, taking a 20-year perspective, WCA sectors have been outperformed on price by Tanzania and Uganda. However, competitive systems clearly also have their weaknesses. Tanzania's competitive sector has been unable to provide any input credit or extension advice, and also performs poorly on lint quality, as predicted. Uganda has maintained a quality premium since liberalization, although this has declined. Since the early 2000s, it attempted to address the challenges of providing input and extension to farmers by curtailing output market competition through the zonal quota system.

To date there is little evidence of the impact of sector structure on research system performance. In theory, companies within monopoly and concentrated systems should have stronger incentives and greater ability to demand high quality research from research systems. However, in practice most research systems remain firmly under state control, such that opportunities for stakeholder influence are limited. The effect of past investment is probably highest in this realm. Mali, with many years of support from French research institutions, was able to carry on some of that momentum, and released six varieties over the past 10 years. However, an emphasis on improved ginning outturn instead of higher seed cotton yields may have reduced the contribution that research could otherwise have made to reversing the yield stagnation seen in WCA countries. Zimbabwe, which featured close collaboration for many years between its own national monopoly, the research institute, and commercial farmers, has also been able to maintain some capacity in this area and released at least four varieties since the mid-1990s. Other countries (with less historical investment in research) have been less successful in this regard, though Zambia has been effective at purifying existing varieties and exploiting more of their yield potential.

Finally, there are few reasons why the performance of seed processing industries should be related to the type of cotton sector organization, other than where the two industries are strongly vertically linked (as historically across West Africa). Indeed, the main finding in relation to seed pricing is that, in the two monopoly systems in which cotton companies also face monopsonistic cottonseed oil markets (Burkina Faso and Mali), prices for seed during 2006 were extremely low. Elsewhere, factors beyond the control of cotton companies (whether a country is landlocked, strength of local demand for cake from the livestock sector, trade policy) largely influence outcomes. More research is needed to understand the performance of African oil and cake markets in greater depth.

## PERFORMANCE ON GLOBAL OUTCOMES: THE OUTCOME INDICATORS

The typology generated clear expectations about one intermediate outcome: that company efficiency would be worst in monopolies and best in competitive sectors. This expectation was strongly confirmed. Adjusted farm gate to FOT costs in Tanzania's competitive sector (the best performer) are less than half those in Mali's national monopoly (the worst performer). In all national and local monopolies, costs are near or above the highest costs in market-based systems, and within these market-based systems, concentrated sectors show higher costs than competitive sectors (such as Tanzania).

The combination of high farmer prices since the beginning of the century and relatively inefficient companies—that is, companies with high operating costs—means that the WCA monopolies are, by a substantial margin, currently the least internationally competitive sectors in the study. FOT cost to value ratios in WCA range from 0.98 to 1.15, compared with a range of 0.76 to 0.88 in all other countries except Uganda. In Uganda, the hybrid regulatory system has kept a large number of ginners in the sector without increasing production, leading to a cost to value ratio of 0.93, the worst in East and Southern Africa (ESA). Other ESA sectors perform well on this measure, but for different reasons. Efficient ginning operations are an important part of the story in Tanzania's competitive sector and Zimbabwe's concentrated sector. The lint quality premiums obtained in Zambia and, to a lesser extent, Zimbabwe boost their performance. However, in both these countries and also in Mozambique, the competitive ratios have been achieved in part because of the low prices they pay their farmers. In Mozambique (which scores 0.79 by this measure) the seed cotton price has been 20 percent to 30 percent lower than the prices in all other ESA countries. In Zambia (the most "internationally competitive" sector in the study, at 0.76), the seed cotton price is substantially higher than it is in Mozambique and not far below the price in Tanzania in absolute terms, but reflects little of the substantial quality premium that Zambian companies receive on the international market.

Returns to farmers have been the best in WCA monopolies, plus Zimbabwe. In WCA these high returns are, of course, heavily influenced by the high prices paid in recent years, which have been coming down and must come down further if these sectors are to become more competitive and sustainable. However, the main reason for the high returns to farmers in WCA is that, by facilitating access to animal traction, fertilizer, and training over many years, these countries have been able to move large shares of farmers into high yielding groups: an average of over 70 percent of WCA farmers are medium-high yield performers (see chapter 10), compared with a range of 20 percent to 30 percent in all other countries except Zimbabwe. Mozambique, a poorly regulated local monopoly, shows the lowest returns, driven by relatively poor yields and very low farm prices.

The picture regarding value added is complex. At farm level, the monopoly systems of WCA perform best on value added per hectare as a result of their high yields. However, a mixed picture emerges in value added per kg of lint because of the high input costs incurred in WCA systems. More efficient ginners in ESA score highly on value added at ginning level. Thus, while

121

Burkina Faso achieved the highest aggregate value added in the year for which these calculations were undertaken, in an unexpected result, three ESA sectors (Tanzania, Zambia, and Zimbabwe) outperformed Cameroon and Mali.[84] Translating these figures into value added per capita (across the whole population), Burkina Faso again performs best. However, the concentrated systems in ESA again perform well, driven by the financial capacity of firms to provide adequate input packages and some extension to large numbers of farmers. Tanzania has the largest population in the sample, so even with its record production in the year in question, its per capita indicator is low. Perhaps the main lesson to be drawn from this analysis is that the relative inefficiency of WCA ginning operations greatly reduces their value added contribution to the wider economy, despite their ability to assist large numbers of farmers in achieving relatively high yields.

Finally, the positive performance on per capita value added in Burkina Faso and Mali has come at a steep cost to the rest of the economy, especially to the state budget, particularly in recent years. Following the 2006 season, Mali's cotton sector required a net budgetary transfer of US$2.47 per capita (US$29 million total; see table 11.5) to cover its losses, while Burkina Faso's required US$0.81 per capita (US$11 million total). Alone among the WCA countries, Cameroon's SODECOTON was able to cover recent losses through surpluses generated in earlier years. Overall, and even excluding Burkina Faso and Mali, the market-based sectors (Tanzania, Zambia, and Zimbabwe) made net per capita budgetary contributions in 2006 at least twice as large as the monopoly or hybrid systems (Cameroon, Mozambique, and Uganda).

In summary, the WCA national monopoly model has generated strong returns to very large numbers of farmers, but poor incentives for cost efficiency have undermined their international competitiveness and their contribution to the wider economy. It is clear that the appreciation of the euro versus the dollar in recent years has contributed to the current lack of competitiveness, but poor cost control within the parastatal companies is also significant. Cost reductions are needed, particularly in Mali, but also to a lesser extent in Burkina Faso and Cameroon (table 11.1). These cost reductions seem unlikely to come without fundamental change in the systems. To accomplish change, policy makers and stakeholders should look at the full range of options, both institutional and technological, at field, ginning, and cotton seed processing levels. A major lesson since 2000 is that the producer price cannot be treated any longer as the main mechanism for ensuring good returns to farmers without jeopardizing the sector's financial sustainability.

Competitive sectors are cost efficient and pay attractive prices to farmers, but their inability to provide input credit and extension, or to raise quality, limits their likely contribution to poverty reduction as long as input and credit market failure remain prominent features of rural Africa. It seems likely that Tanzania's competitive system has been able to perform as well as it has in part because of favorable agro-ecological and population settlement characteristics. However, a competitive model's performance could be expected to be substantially poorer in many areas of WCA, which are less well endowed in cultivable land and soil fertility.

Concentrated sectors have performed well on a broad range of indicators. They have scored highly on quality and service delivery (input and extension), have been more efficient than the monopolies, and have also generated attractive value added per capita while making the highest contributions to state budgets through taxes and fees. Yet, since 2000 their performance on seed cotton pricing has been disappointing. As illustrated by the problems caused by new entry in Zimbabwe since 2001 and in Zambia since 2004, the long-term performance of these systems is likely to depend on a supportive approach to regulation, something that has yet to be achieved within an African cotton sector (see below for more details on key elements of such regulation).

## Ways Forward for African Cotton: Cross-Cutting Challenges

In assessing the performance of African cotton sectors, it must be kept in mind that, while some are faring better than others, on a global scale all are lagging behind the best performers in the world on one or several dimensions. Though they have expanded their share of the world lint market since 1970 these sectors face increasing competition, hence low future prices. Competition means that they must continually cut costs, raise productivity, and add value if they are to maintain attractive returns to farmers and to make a positive contribution to national poverty reduction goals. To achieve these purposes, all African cotton sectors need to improve their performance on critical issues such as improved research and extension responsiveness and efficiency, technology transfer in areas such as dissemination of genetically modified varieties, lint quality management and marketing, soil conservation, and technical support to farmers and farmer organizations. Effective strategies for African cotton sectors should, therefore, combine institutional innovations and reforms with necessary additional investments in key public goods.

This section considers various cross-cutting actions that African cotton sectors should take to build on comparative advantages and narrow existing performance gaps with international competitors. These efforts can be articulated around the following three major objectives: (i) achieving greater value through improved quality, marketing, and valorization of by-products; (ii) bridging competitiveness gaps through farm-level productivity and ginning efficiency; and (iii) improving the sector's sustainability through institutional development and capacity building of stakeholders, as well as strengthening of governance and regulatory structures and management systems. Some of the actions discussed below could usefully be tackled at a regional level, as well as nationally, and may benefit from donor support.

### ACHIEVING GREATER VALUE

Efforts to achieve greater value in SSA cotton sectors should focus on three priority areas: quality, marketing, and valorization of by-products.

*Quality.* African cotton has two potential competitive advantages in the world market: the intrinsic quality of its fiber (the fiber properties) and the fact that it is handpicked. Quality improvement—especially the elimination of contamination—could therefore result in selling prices of up to US$0.10 per pound above the Cotlook A Index. At typical producer prices of US$0.25–US$0.32 per kg, a US$0.10 per lb increase in the price of lint that is fully passed on to farmers would increase farmer prices by 30 to 40 percent. As a result, quality management should be considered one of the most important areas of improvement for SSA cotton exporting countries.

Though most African cotton is suitable for the medium-high level of ring spinning, progress on quality since the mid-1990s has generally been disappointing and certainly not as fast as required by the spinning industry. The trend in spinning technology toward more automation and higher speeds makes improvements in quality and consistency a vital issue for the future of African cotton sectors. Yet the quality reputation of many African lints has been eroded, primarily by

123

contamination from foreign matter. Meanwhile, the impact of quality on cotton pricing is not fully understood by producers, and even by some smaller ginners.

Greater awareness may be necessary to reestablish Africa's main comparative advantage stemming from the manual harvesting of seed cotton. However, Zimbabwe's experience since 2003 clearly illustrates that awareness alone is not enough. The Zimbabwe sector enjoyed a strong quality reputation on international markets, which the dominant firms were keen to maintain, and farmers were disciplined in grading their seed cotton before sale. However, when new entrants began offering flat-rate prices irrespective of quality in 2003, farmer practice changed within a season and the average quality of seed cotton delivered to buying points plummeted. Indeed, a central insight from this study is that sector structure has an impact on both the incentives that ginners face to achieve high quality lint and on their ability to control their supply chain to achieve it. Technical solutions to eliminate contamination are well known, but some form of coordination across firms is necessary if such solutions are to be implemented effectively.[85]

Quality improvement in SSA requires a concerted effort from researchers, farmers, and ginners, to improve fiber characteristics through research and better production practices, reduce variability of lint quality through more rigorous seed cotton grading and lint classification, control contamination through capacity building and price incentives, and optimize quality management in ginning. Efforts must also be made to generalize the use of cotton cloth wrappers for bales and, ultimately, also to develop container loading at the gins and optimize export logistics.

However, the typology developed in chapter 4 highlights a key conundrum in African cotton systems: while concentrated (and perhaps monopoly) sectors are more likely to achieve the coordination needed to improve quality, they are not necessarily ready to pass the resulting price premiums on to farmers. For example, the comparative analysis in previous chapters showed that farmers in Tanzania have received slightly higher prices than farmers in Zambia, despite the much higher price premium in Zambia. Solving this riddle—how to capture the very significant price premiums available to African cotton while sharing some of that benefit with farmers—probably requires much stronger farmer organizations than are currently found in SSA. In concentrated systems, these organizations would have to bargain with the large ginners for remunerative prices on the basis of solid knowledge of world prices, realized export prices, quality premiums obtained, and cost structure borne by ginners. In competitive systems these organizations would need to focus on training farmers about the benefits of increased quality, and on monitoring prices paid by companies to ensure transmission of quality premiums.[86]

*Marketing practices.* Discussion in chapters 2 and 7 highlighted the fact that African cotton sectors commonly lose revenue from weaknesses in lint marketing, not just lint quality. Some weaknesses are largely beyond the control of the cotton sector, especially where a landlocked country relies on the export infrastructure of a neighbor. African port facilities are often inefficient and shipping unreliable, while the poor state of national road networks contributes to high internal transport costs and delays. As a major export industry in many African countries, the cotton sector should be lobbying hard for improved infrastructure. This may also be an area where international development finance can assist.

African sectors are also progressing slowly with the introduction of high volume instrument testing of lint. Use of instrument testing on a bale-by-bale basis requires purchasing the expensive equipment, training scientists, and equipping laboratories to provide reliable results from such equipment under African conditions. Introduction of this equipment will require increased international technical assistance beyond current levels.

With regard to marketing practice at firm level, cotton ginning companies affiliated with international merchants (referred to in this book as "affiliated ginners") have better access to market information and also benefit from hedging against price and exchange rate risk undertaken by their parent companies. Among independent ginners, larger firms such as the parastatal companies of WCA, and Cottco in Zimbabwe, are better placed than smaller, often also newer, rivals to offer large volumes of particular quality lint or year-round sales, or both. They should also possess greater accumulated knowledge of international markets and be better placed to bargain with international merchants.[87]

Many options exist for independent ginners in SSA to improve marketing performance. Forward sales, the most common marketing method in the cotton business, are the easiest and most effective marketing strategy to cover risks. The flexibility and effectiveness of such sales can be enhanced if they are supplemented by the use of market instruments such as futures and options; the application of these instruments can be specified in the physical contract arrangements with merchants. Ginners can also spread their risks by committing portions of the total production to different marketing options: cash sales after ginning and other options requiring commitment before harvest; forward sales at fixed price; "on call" at price-to-be-fixed contracts; and minimum guaranteed price contracts. Direct sales from ginners to spinners through commissioned agents can save the cost of intermediation by merchants while improving the quantity and quality of market knowledge available to the ginners. A final potential tool for improving the marketing of African cotton is electronic trading. E-trade platforms can be a very effective means of transparent price discovery brought about by real-time multilateral bids and offers, and online contracting can reduce transaction costs.

To realize improvements such as those recommended above, independent ginners in SSA need to be informed and trained to better understand the world cotton market and prices, master cotton trade rules and regulations, and understand how to use risk management techniques based on futures and options contracts. There is a potential role for international organizations in this area.

Finally, stronger farmer organizations can also work to the advantage of the sector if it allows more systematic contractual trade relationships between farmers and ginners. Examples could include more formalized contract farming relationships than currently exist, with such contracts specifying predetermined volumes of seed cotton to trade, with precise quality specifications. Stronger farmer organizations might also be better informed about world markets and be better able to negotiate a pricing approach that is tied to world prices, including recognition of world market premiums for quality. Considerable institutional strengthening and training will be required first to reinforce producer organizations.

***Valorization of seed cotton by-products.*** As emphasized in previous chapters, the performance of the oil and cake sectors (which affects the price that ginners receive for their seed and, ultimately, the price that farmers might receive for their seed cotton) requires more attention than it has received so far. Internationally, these markets are changing fast as a result of the increased demand for edible oil and animal feed related to competitive pressures on alternative uses of cereals and raw materials for biofuel production. Meanwhile, intra-Africa and international comparisons indicate that some WCA sectors, in particular, receive very low prices for their seed, although others do much better. Improving the valorization of cotton seed has a number of important policy implications, and can take different forms:

- Movement toward open and transparent oil markets may bring about better prices for the seeds and better outcomes through increased competition (such as improved investment climate, selection of professional investors, stronger enforcement of the rules and regulations on imported oil, and so forth).

- Focused efforts to develop strategies with stakeholders for cottonseed oil and cake are also important. Despite the economic weight and strategic importance of these activities, few countries have developed a strategy for cotton seed industries; this is an area donors might want to consider supporting at the domestic, subregional, and international market levels.

## IMPROVING THE PRODUCTIVITY AND COMPETITIVENESS OF COTTON PRODUCTION IN AFRICA

Bridging productivity and competitiveness gaps is also critical for strengthening African cotton sectors. The main areas of focus for these efforts are productivity at the farm level and efficiency of ginning industries.

***Productivity of cotton cultivation at farm level.*** Increasing productivity at farm level is necessary to improve a sector's overall competitiveness and to make the cotton crop more profitable for farmers. As shown in chapter 10, average yields in WCA sectors are about at the world rainfed average, but have stagnated since the mid-1980s, while rainfed cotton yields in other parts of the world have risen rapidly. Average yields in ESA countries have shown some improvement over time,[88] but remain much lower, and indeed have risen more slowly, than the world rainfed average. Delivery of high quality input packages to producers is clearly one requirement if productivity at farm level is to be increased. The analysis in this book has shown that some sector types (monopolies and concentrated sectors) are much better at input delivery than others. However, there are also several areas where African sectors exhibit more general weaknesses. These areas, often linked to the provision of public goods within a sector, are points of possible external technical or financial intervention.

***Improvements in the delivery of extension services and technical assistance.*** This study has shown that some sector types can provide reasonable, basic extension services (covering simple but fundamental agronomic messages and perhaps support for animal traction investment) to farmers. Traditional single-channel systems, which directly or indirectly provide extension services to all cotton growers, delivered remarkable results until the mid-1980s. However, such systems have since exhibited decreasing effectiveness and efficiency as the cotton companies concerned have grown larger, have not been under any competitive pressure, and in some cases, have become more politicized. Concentrated sectors in Zambia and Zimbabwe have delivered reasonable extension support to farmers, while local monopolies can (but do not necessarily) invest in extension. The Zambian example is instructive—donor assistance was used to dramatically scale-up the level of technical support provided by the main company.[89] In contrast, competitive systems are struggling to deliver extension services at all. With competitive systems, even the provision of basic agronomic messages may require some degree of public support (from local governments or donors, ideally in conjunction with both the cotton board and ginners' association).

In all systems there are also wider issues that the cotton sector alone is unlikely to be able to address adequately, although it will be a beneficiary when progress is made. Soil fertility management is a long-term challenge, with benefits that extend well beyond cotton. Promotion of safe chemical application and integrated pest management are also sufficiently long-term endeavors that they are likely to suffer from free-riding by cotton companies, except those in secure monopoly arrangements.[90] Local monopoly systems may also be the only ones in which companies have adequate incentives to invest in the development of locally adapted fertilizer recommendations (a particular issue in WCA, given the high share of gross cotton revenue that is expended on input). In addition, particularly now that former parastatals have been freed from their responsibilities to promote broadly based rural development in cotton areas, cotton companies may find it too costly to support animal traction investment by farmers other than the medium-high performing ones (who already deliver quantities of seed cotton that make them creditworthy and able to pay off lumpy investments in reasonable time). For poorer farmers, acquisition of animal traction assets is likely to be part of a longer-term process of asset accumulation (including better soil fertility management) in which livestock of various forms play a central role. This subject links to a final area in which cotton companies have few, if any, incentives to contribute: developing alternative (agricultural and nonagricultural) income-earning opportunities for households in cotton areas that are unlikely to benefit from any of the above. In all of these areas, some public role beyond the cotton industry (national extension or local government program, with possible donor support) can be envisaged. In most of these cases, however, working with the cotton industry will enhance the efficiency and effectiveness of the efforts undertaken.

*Improvements in research.* Publicly funded cotton research in Africa appears weak but nevertheless has a vital role to play in helping to ensure the competitiveness and sustainability of the continent's cotton sectors over the long run. This claim is supported by three arguments. First, many (possibly the majority) of Africa's cotton farmers are, and will likely remain, too resource-constrained to close the yield gap with existing technology through more effective and timely management of weed and insect pests on both their food and cotton crops. Second, if cotton prices continue to decline as a result of productivity increases in other cotton producing areas of the world, or real food crop prices increase as a result of global demand for plant-based energy feedstocks (a trend that will be aggravated further by migration from rural areas), farmers will find cotton less and less attractive over time with current technology. Finally, soil fertility management has become a critical issue in several West African countries, and is also affecting even the better endowed soils of ESA, as evidenced by high levels of striga infestation. Improved cotton production technology therefore needs to be embedded in sustainable cropping systems to be socially and economically viable in the long run.

Improvements in research performance will depend on strengthening internal and external links between researchers and other stakeholders. External links across research organizations are necessary to achieve critical mass, given the small size of individual national programs, and to maximize potential spillovers among researchers addressing common technology constraints and opportunities. Currently, no formal cotton research networks exist in ESA and those in WCA are in urgent need of rejuvenation (the West and Central African Council for Agricultural Research and Development [CORAF/WECARD], for example). This is a possible area for donor action. With regard to internal links, irrespective of sector type, a high policy priority should be to move toward greater involvement of ginners and farmers in research management. This approach should institutionalize stakeholder involvement in setting research priorities, monitoring research performance, and accounting for how research funds have been used, thereby permitting greater funding of research efforts by the industry through either direct contributions or levies. Ideally, it should also allow stakeholders, through whatever management regime is put in place, to appoint and fire researchers and determine their salary scales, rather than relying on public sector practices and scales.

*Technology transfer.* A major technology that is likely to be of particular interest to cotton growers in Africa over the near to medium term is genetically modified cotton.[91] Currently, Bt cotton is by far the most common genetically modified type and the most relevant for Africa, but additional innovations already exist (herbicide resistant, "stacked" genes) and more will certainly be developed. Bt cotton has undoubtedly been a major source of yield gains in cotton in India (much of it rainfed) since 2000. While there are technical, organizational, and public policy challenges that need to be addressed if Bt cotton is to be introduced into African cotton systems, there is every reason to believe that Bt cotton varieties would generate significant productivity gains in African countries where pest control is poor, and should be able to reduce marginal production costs where chemical input use is higher and pest control is more effective.[92] The key factor influencing profitability in both cases would be the licensing fee that has to be paid for the technology. Among the countries studied here, currently only Burkina Faso has completed testing of Bt cotton varieties. (The country is now in the negotiation phase with Monsanto over patent issues, including licensing fees.) Most African countries have yet to begin testing Bt cotton varieties to evaluate the potential gains, and some even have a moratorium on such testing. Clearly, there needs to be a public debate in many countries before Bt cotton can be commercially released. However, there are also important technical steps to be taken, including the development of biosafety assessment procedures (where lacking) and equipping of laboratories, and breeding work to incorporate the Bt gene into locally adapted cotton varieties. There may be regional economies of scale in aspects of this work, hence the World Bank biosafety project now being implemented in several WCA countries; further assistance from international agencies could be useful in this area.

*Efficiency of ginning industries.* The efficiency of ginning industries is critical to the competitiveness and sustainability of the sector overall. In competitive and concentrated systems, companies demonstrate significantly lower operating and overhead costs than in monopoly systems. In national or local monopolies there are too few incentives to improve performance and efficiency of operations and to reduce costs. Achieving significant productivity gains at this level may imply structural changes well beyond the usual—and so far fairly unproductive—pressure put on these monopolies:

- Reducing the costs of ginning and other postharvest activities may entail revisiting policies on the choice of technology (for ginning and cotton seed processing), the size of the industrial units, the fiscal incentives to minimize investment costs by allowing secondhand equipment to be imported, and the profile of strategic investors. As mentioned previously, these factors are strongly linked to the sector's organizational model.

- Helping develop real cotton industry clusters—where related services (maintenance, transport, financial services) and input can be procured at competitive prices—could also contribute to improving the performance and costs in cotton sectors.

## IMPROVING SUSTAINABILITY, GOVERNANCE, AND MANAGEMENT OF COTTON SECTORS

The financial sustainability of cotton sectors is very much linked to sector organization. The traditional single-channel systems of WCA (especially Benin, Burkina Faso, and Mali) have experienced severe and recurrent financial crises, as much a result of the lack of adjustment capacities of these systems as of world cotton price fluctuations and changes in the dollar-euro exchange rate. By contrast, competitive and concentrated systems in ESA have been operating without requiring public subsidies or creating fiscal liabilities since the liberalization of these sectors in the early 1990s. In WCA, more realistic price-setting mechanisms, improved risk management techniques, and new marketing strategies can help mitigate such financial problems in the future. However, greater adaptability is also likely to require a change in business culture and attitudes within the cotton chain, which may only come from greater involvement of the private sector through the entry of national and international professional operators with long-term commitments to improving sector performance. Examples of areas in which business culture and attitudes (among companies and farmers) could usefully change include quality control and input access, with an evolution from farmers' rights and state provision toward commercial transactions. For input access, this would mean a movement toward differential input access according to a farmer's ability to make productive use of input, in contrast to the current standardized provision irrespective of production capability.

Improving sector management and governance should be high on the agenda of all African cotton sectors, although specific governance needs vary by sector type. In monopoly systems (and perhaps in concentrated systems in the future) building inter-professional committees that can effectively and wisely perform their price-setting and other tasks, is a high priority. Specifically in local monopoly systems, definition and enforcement of clear rules for evaluating and re-tendering concession areas (that is, not just for allocating them initially) is a key task. Mozambique's failure to re-tender concessions after almost 20 years of private sector involvement lies at the root of its disappointing performance in many areas. In concentrated systems, similar priority should be attached to the development and implementation of licensing criteria that set out clearly the capabilities and conduct required if a firm is to participate in the sector. In Zimbabwe, draft regulations of this nature have awaited official ratification for four years. Meanwhile, in competitive sectors the need for the government to work with stakeholders to play a central coordinating role to ensure that farmers can access input and technical advice carries with it the requirement that such a role be performed as efficiently and transparently as possible, taking into account the views of all main stakeholder groups. Finally, as already noted, reform of the management and governance of research organizations to make them more responsive to other stakeholders is a priority in almost all sectors.

Just as improved sector management and governance are desirable, irrespective of sector type, so farmer organizations should be strengthened in all sector types to enable them to play a more effective role in sector governance and in service delivery. Strengthening farmer organizations is an area where donor funding is particularly needed. States may be reluctant to fund the strengthening of civil society organizations and, if they do so, may have an agenda of political control rather than true capacity building. In turn, however, donor-controlled projects may not be the best vehicle for capacity-building activities. Instead, program implementation could be contracted out to nongovernmental organizations with track records in supporting farmer organizations.

## REFLECTIONS ON INVESTOR PROFILES: GLOBAL AND LOCAL CHAMPIONS

A final issue of concern to policy makers in all sectors—and one that intersects with several of the points raised above—is the type of firms that should be encouraged to invest in the industry.[93] Large firms have a number of advantages over small rivals. They have more resources to invest in preharvest service provision and can afford to hire specialist technical expertise in critical areas (for example, agronomy, ginnery management, lint marketing) because there are economies of scale in utilizing such knowledge within organizations. When larger volumes of lint are produced, the opportunities for forward contracting, year-round sales, and large consignments increase, as does bargaining power with international merchants. If the large firm is itself affiliated with an international merchant, additional benefits flow from the unrivalled access to market information and a pool of technical expertise, and from the price and exchange rate hedging that such merchants undertake.

Investment by international merchants in a national cotton sector is generally, therefore, a good thing. Compare the performance of the newer concession areas in Mozambique (many taken by affiliated ginners) with that of the earlier concessions, or witness the leading role played by Dunavant in promoting the Zambian sector. However, this does not mean that countries need depend entirely on such firms to drive sectoral development. In Zimbabwe, Cottco provides an excellent example of a national champion that has maintained a dominant position within the cotton sector in the face of international competition and, but for the recent economic crisis in Zimbabwe, would probably have established itself as a major force in Uganda and Mozambique. Arguably the key to Cottco's success has been its combination of accumulated technical expertise from its highly competent parastatal predecessor and the private sector entrepreneurship and management unleashed by privatization. In Burkina Faso and perhaps in other WCA sectors, wise privatization of the current parastatal cotton companies (probably combined with breaking them up into more manageable units, as occurred in Zambia when Lintco was privatized) would produce similar national champions. Having such firms competing with international merchants in domestic markets—and perhaps in due course in regional markets—will increase the political acceptability of sector reform and also provide some guarantee of long-term commitment to development of the national industry.

This then leaves the question: is there any role for smaller firms in a high-performing cotton sector? Tanzania's sector will continue to be based on such small firms for the foreseeable future. However, even in concentrated sectors and perhaps in local monopolies (if there are one or two small concession areas), small firms can use their lower overhead to keep big firms focusing on their own efficiency and on paying reasonable prices to farmers. Moreover, small firms are the only feasible entry point for innovative local entrepreneurs who wish to enter the cotton industry and have to start somewhere. A key lesson from this study, however, is that such firms must be made to adhere to strict codes of conduct if their presence within a sector is to do more good than harm.

# Ways Forward for Particular Sector Types

While there are some common challenges across African cotton sectors, many challenges are more acute in some sector types than others. Moreover, the appropriate response to most challenges will depend heavily on the type of sector in question. Thus, this final section considers the key challenges and opportunities facing each of the main sectoral types. Some reflections on future trajectories for cotton sector organization in Africa are given as a preface.

## FUTURE TRAJECTORIES FOR COTTON SECTOR ORGANIZATION IN AFRICA

The typology presented in this book offers a strong and reliable framework to provide insights into possible evolutionary paths of African cotton sectors and to guide decision makers on the possible paths of future reform. In the short to medium term, the most likely change within African cotton systems is an increase in the number of local monopoly systems in WCA. However, local monopolies should be a transition phase toward market-based sector types such as concentrated and competitive systems. If so, the most desirable end type is probably a concentrated system, which has a wide range of desirable properties if regulatory challenges can be overcome to make them more stable. For example, if clear licensing rules can be developed, total regulatory costs under a concentrated system may be lower than with local monopolies,[94] while incentives for cost reduction are greater. Assuming that appropriate regulatory models can be developed, the strength of farmer associations in WCA (relative to most countries of ESA) could mean that the local monopoly stage could be a reasonably short one.[95] However, additional attention must still be paid to educating farmer associations about the realities of the world cotton market and to increasing their operational capacities.

More-competitive systems are perhaps the long-term future. However, there is a need for stronger farmer associations to take over some critical functions (for example, extension) and for improvements in rural input and financial markets before competitive systems can support genuinely high-performing cotton sectors in most countries. For the foreseeable future, competitive systems will have a hard time increasing productivity and quality to such an extent that they make a major contribution to reducing poverty. Therefore, stakeholders, policy makers, and donors cannot avoid dealing with the details of institutional design to cope with input and credit market failures. This design needs to be tailored to the current market structure and historical patterns of the country in question.

Overall, one may expect some degree of convergence in the forms of cotton sector organization seen in SSA over the next decade, with emphasis on a degree of private sector competition, an important role for farmer associations, and a multistakeholder approach to sector regulation. This convergence should be accompanied by policies and programs aiming, across sector types, at improving the quality and marketing of cotton lint, reforming and improving research and research-extension links to close the productivity gap, and strengthening institutional capacity at all levels. Effective strategies for African cotton sectors should therefore combine the necessary institutional reforms—given that this book has demonstrated a generally positive response to reforms so far—with a set of other coordinated actions (with donor support) to "raise the game" on some critical issues to be handled at national and regional levels, such as technology, soil conservation, or technical support to farmers and farmer organizations.

As suggested in the typology, change in national monopolies depends on policy choice. Cameroon, with relatively good performance to date, is an example of a country that may be able to maintain its national monopoly to good effect, as long as it reforms its price-setting process. The creeping inefficiencies identified in this book, however, seem likely to force change even in Cameroon at some point. History and accumulated experience in WCA (path dependency) suggest that initial change in the region will predominantly be toward local monopolies, and recent policy decisions in Burkina Faso and Mali support this conclusion. In the context of a possible continued high dollar-euro exchange rate, the biggest challenges in these sectors include improving cost effectiveness of the cotton companies, tackling stagnant productivity at farm level, raising quality, developing pricing formulas that make price setting more connected and responsive to world market prices, and identifying ways for cotton companies to improve management of the intraseasonal price risk that they incur because of the panterritorial, panseasonal, price-setting mechanism. What, then, must these countries do to tackle these problems in a local monopoly setting?

To begin to answer that question, one could turn to the experience of Mozambique, which has shown that performance under such systems can be quite poor. Yet this country is probably an inferior predictor of the performance of the WCA systems for several reasons. First, Mozambique at the time of reform had almost none of WCA's history of substantial farm-level input use, investment in animal traction, technical advice to farmers, and regular release of new varieties. WCA thus starts at a much higher level than Mozambique and needs to renew productivity growth rather than start it from zero. Second, the inter-professional approach to sector coordination that has emerged in most WCA countries provides much greater promise of consensual sector management than has been observed until recently in Mozambique. Finally, and related to the last point, farmer organizations are more developed in WCA than in Mozambique, though their technical capacities remain uneven and require further strengthening.

Recent positive developments in new concession areas of Mozambique do hold a lesson for WCA. First, private sector capital and management must have a prominent role in the reformed sectors. Second, not all private capital will perform well. Policy makers need to choose private investors carefully to ensure that they have the technical and managerial knowledge, long-term commitment, and financial capacity to deliver high quality services to farmers.

If WCA countries do move to local monopoly arrangements, key factors they need to take into account include the following:

- Cost reduction from farm gate to FOT needs to be a top priority. To reduce costs, private companies need a greater role in price setting and other decision making than they have so far been given in Burkina Faso. The fact that SOFITEX controls 85 percent of the market, combined with its apparently soft budget constraint and the sector-wide price-setting mechanism, has done little if anything to spur sector-wide cost reductions. If the government is to maintain a role in a cotton company, it must do so at a substantially lower market share.

- If concession zones are to be auctioned, care must be taken to avoid sales prices so high that they make it difficult for the new companies to compete. In Burkina Faso, high

132

auction prices may have undermined the new companies' ability to compete with SOFITEX in a context of declining world prices for lint and overvaluation of the currency.

- Price setting needs to occur in a framework of negotiation, but rules must continue to be reformed to provide reasonable assurance to companies that, if they operate efficiently according to international standards, they will be able to earn a reasonable return on their investment over time. Some level of price flexibility over the course of the marketing season may need to be a part of the revised pricing approaches. Cotton companies need to recognize and improve the management of their exposure to cotton price and exchange rate volatility.

- Inter-professional committees and farmer organizations need to continue to be developed, with special emphasis on the operational abilities of the latter.

- Reforms in research organizations continue to be needed to make sure that they are responsive to these inter-professional committees.

- Clear rules for evaluating and re-tendering concession areas need to be developed, as this has been a key failure in Mozambique.

- To create more competition, investment and structural reforms in the cottonseed oil sector should be encouraged.

Three findings from this research are especially relevant if WCA countries instead consider moving to a concentrated, market-based system. First, the systems in Zimbabwe and Zambia have suffered periodic bouts of instability. Zimbabwe, in particular, appears to have crossed a tipping point since 2001, whereby the entry of additional ginning companies has undermined existing mechanisms for coordination of input supply, extension, and quality control. Regulation of concentrated systems is thus a key challenge (see below). Second, WCA's agro-ecological conditions (especially the low fertility of its soils) suggest that a competitive sector (which could be the outcome of instability within a concentrated sector) may perform quite poorly unless farmer organizations themselves are strong enough to ensure broadly based access to input. Finally, though farmer organizations in most WCA countries are much stronger than in ESA countries, few if any appear strong enough to take on this challenge in the near future. Moving to a fully privatized market that allows competition among companies, even if the market is initially very concentrated, is thus a risky proposition for WCA countries. If instead these sectors can use the local monopoly approach to build up the operational capacity of farmer organizations and develop sound regulatory mechanisms concentrated and eventually competitive systems could perform well.

Meanwhile, the key challenge in Mozambique's local monopoly sector is how to create incentives for good company performance within the concessions. In the absence of strong farmer associations, these incentives have to come from some combination of improved rules governing tendering and re-tendering concessions, procedures for monitoring performance of concessionaires, and careful selection of companies. It appears that Mozambique has done a good job on the latter, with newer companies clearly outperforming original concession holders; the country is also developing serious proposals for evaluating and re-awarding concession areas. Prices to farmers remain very low, however, and are unlikely to improve without improved regulation.

## CONCENTRATED SECTORS

Change in concentrated sectors is likely to be driven less by policy choice than by inherent characteristics of these systems. Because investment in ginning capacity is not prohibitively expensive—especially where investors are familiar with roller gin technology—concentrated systems can move, over fairly short periods, toward more competitive systems. Such a development may eventually improve prices to farmers, but can also have negative implications for credit repayment (and therefore future input credit provision) and quality. Zimbabwe has seen problems in both areas since 2003, while Zambia's problems in 2006/07 affected credit repayment and provision, but are too recent to have had observable effects on quality. In both cases, new entrants have so far been too small to exert much price pressure on existing companies. Indeed, the biggest danger with such new entry is that increased competition undermines input credit and lint quality well before it has any positive effect on prices paid to farmers.

The key challenge for concentrated sectors, therefore, is to develop a flexible and commercially supportive regulatory regime that understands the strengths and weaknesses of the concentrated model:

- Concentrated sectors need clear and transparent barriers to entry (licensing rules that specify strict capabilities and conduct of firms wishing to participate in the sector) to defend the ability of firms within the sector to coordinate on input supply, extension, quality control, and perhaps other matters.

- Concentrated sectors must retain some contestability to provide incumbents with an incentive to maintain attractive seed cotton prices. As in the case of local monopolies, it is important for those in charge of policy for the sector to form a clear idea of the type of company that they wish to allow into the sector, so as to be able to formulate rules accordingly. Given the tendency of these sectors to slide toward unrestrained competition and credit default crises, a strong commitment to raising farmer productivity and improving quality within the chain should be given high priority in the selection criteria.

- However, given the problems of relying entirely on the threat of entry to discipline incumbent firms within concentrated sectors, it may also be desirable to develop price-setting mechanisms that are more formalized than the price leadership that has prevailed in concentrated systems thus far. As piloted in WCA sectors, farmer organizations have a potentially important role to play within such mechanisms; but this role needs to be informed by a solid understanding of world markets to avoid the problems seen in WCA.

## COMPETITIVE SECTORS

As could be expected, there are fewer concerns over pricing within competitive sectors than in other sector types. Instead, the weaknesses are in service delivery and quality control. Given the pervasive failures in credit and input markets in rural Africa, the typology suggested that competitive sectors may face pressure to move toward more coordinated sectors and that, if movement were to occur, it would most likely be toward a local monopoly or hybrid system. Uganda began experimenting with solutions to input and seasonal finance market problems nearly as soon as it emerged from reform with about 30 active ginners. In 2003, it moved to a hybrid sector model that included zoning and seed cotton quotas, as a way of providing incentives for investment in input supply. However, this model has also encountered problems and has been suspended for the 2007/08 season. By contrast, Tanzania has developed an approach to input supply that features an important role for the government, but has always incorporated arrangements that allow it to preserve the strong competition among firms in the market for seed cotton.[96]

Key insights from this work regarding competitive systems include the following:

- Such coordination as does occur within a competitive sector must come from some central body. Given the difficulty of obtaining consensus among large numbers of competing ginners, the state is likely to have to play a key role within this body. This is in contrast to local monopoly or concentrated systems, where inter-professional committees dominated by ginners and farmers have more potential to adequately manage the sector. The risk with allowing a state agency to play such a central role is that it can make mistakes, even if well intentioned, and could do worse if rent-seeking or other motivations prevail. Thus, the accountability of regulatory bodies to ginners and farmers needs to be strengthened.

- Incentives within competitive sectors for individual ginners to support long-term programs, such as initiatives to enhance soil fertility or promote animal traction, are extremely limited. Thus, the cotton board and the ginners' association may have to work with other actors (local government or donors, for instance) to develop programs that enhance the asset base of farmers and also generate benefits beyond the cotton sector. In the same way, they could possibly explore seasonal financing models (for example, SACCOs[97] in Tanzania) that, over time, might allow some farmers to access greater quantities of purchased inputs for cotton production.

- Uganda's hybrid approach to solving the input credit problem in competitive systems kept all ginners in the market, but after four years had failed to increase total production despite improvements (though erratic) in service delivery. The system, therefore, entrenched the sector's chronic overcapacity, leading to operating costs that were much higher than they would otherwise have been, especially considering the heavy use of roller gins in the country. Generalizing, we suggest that hybrid approaches within competitive sectors need to avoid protecting ginners entirely from competitive pressure from within the country.

- Tanzania's agro-ecological and population settlement characteristics have so far protected it from the need to take the radical measures that Uganda took for input credit provision. However, if yields begin to fall as a result of declining soil fertility (or possibly increasing pest pressure), and if it wants to more fully realize its potential, the country may need to consider moving to a more coordinated approach.

Uganda's hybrid approach to solving the input credit problem in competitive systems kept all ginners in the market by granting them all quotas for output purchase. However, after four years, the resulting (erratic) improvements in service delivery had failed to stimulate an increase in total seed cotton production. The main consequence of this measure, therefore, was to entrench the sector's chronic overcapacity, leading to operating costs that were much higher than they would otherwise be, especially considering the heavy use of roller gins in the country. Generalizing, we suggest that hybrid approaches within competitive sectors need to avoid protecting ginners entirely from competitive pressure from within the country.

Tanzania's agro-ecological and population settlement characteristics, permitting expansion of "extensive" cotton production, have so far protected it from the need to take the type of radical measures that Uganda took for input credit provision. However, if yields begin to fall due to declining soil fertility (or possibly one day to increasing pest pressure), and if it wants to more fully realize its potential, the country may need to consider moving to a more coordinated approach.

# Appendix A: Statistical Tables

**Table A1 Benin**

| Season | Lint production (thousand tons) | Area (thousand hectares) | Yield (kg lint/hectare) | Ginning ratio (%) | Grower price (CFA f/kg seed cotton) | GDP deflator (2000 = 1.00) |
|---|---|---|---|---|---|---|
| 1970/71 | 14 | 39 | 351 | 38.1 | 34 | 0.16 |
| 1971/72 | 18 | 55 | 333 | 38.4 | 35 | 0.16 |
| 1972/73 | 19 | 48 | 396 | 37.9 | 35 | 0.17 |
| 1973/74 | 17 | 53 | 329 | 38.6 | 37 | 0.17 |
| 1974/75 | 13 | 49 | 256 | 40.5 | 45 | 0.20 |
| 1975/76 | 8 | 32 | 248 | 39.5 | 45 | 0.23 |
| 1976/77 | 7 | 26 | 260 | 38.7 | 50 | 0.26 |
| 1977/78 | 5 | 21 | 249 | 37.7 | 55 | 0.27 |
| 1978/79 | 7 | 26 | 275 | 38.2 | 55 | 0.31 |
| 1979/80 | 10 | 26 | 372 | 37.6 | 55 | 0.35 |
| 1980/81 | 5 | 30 | 167 | 37.8 | 60 | 0.38 |
| 1981/82 | 5 | 24 | 228 | 37.9 | 80 | 0.41 |
| 1982/83 | 12 | 24 | 490 | 37.9 | 85 | 0.48 |
| 1983/84 | 17 | 40 | 430 | 37.4 | 100 | 0.50 |
| 1984/85 | 33 | 56 | 597 | 37.8 | 100 | 0.51 |
| 1985/86 | 34 | 100 | 338 | 38.0 | 110 | 0.49 |
| 1986/87 | 48 | 103 | 464 | 39.0 | 110 | 0.47 |
| 1987/88 | 27 | 72 | 380 | 38.9 | 100 | 0.48 |
| 1988/89 | 44 | 97 | 456 | 40.5 | 105 | 0.48 |
| 1989/90 | 43 | 111 | 383 | 40.7 | 95 | 0.49 |
| 1990/91 | 59 | 123 | 482 | 41.2 | 100 | 0.50 |
| 1991/92 | 75 | 144 | 518 | 42.2 | 100 | 0.50 |
| 1992/93 | 69 | 139 | 493 | 42.5 | 100 | 0.52 |
| 1993/94 | 103 | 235 | 439 | 41.9 | 110 | 0.53 |
| 1994/95 | 98 | 230 | 426 | 41.9 | 140 | 0.70 |
| 1995/96 | 141 | 294 | 481 | 40.5 | 180 | 0.81 |
| 1996/97 | 143 | 292 | 491 | 41.2 | 200 | 0.86 |
| 1997/98 | 150 | 386 | 389 | 41.8 | 200 | 0.91 |
| 1998/99 | 138 | 394 | 351 | 41.3 | 225 | 0.95 |
| 1999/2000 | 152 | 372 | 409 | 41.9 | 185 | 0.97 |
| 2000/01 | 141 | 337 | 418 | 41.5 | 200 | 1.00 |
| 2001/02 | 172 | 357 | 482 | 42.1 | 200 | 1.03 |
| 2002/03 | 143 | 313 | 457 | 42.4 | 185 | 1.05 |
| 2003/04 | 142 | 323 | 440 | 42.5 | 205 | 1.11 |
| 2004/05 | 171 | 325 | 527 | 41.8 | 190 | 1.08 |
| 2005/06 | 82 | 200 | 408 | 41.8 | 185 | 1.14 |
| 2006/07 | 103 | 236 | 438 | 42.0 | 170 | — |

*Source:* SONAPRA for cotton statistics; IMF International Financial Statistics and World Bank estimates for GDP deflator.

*Note:* CFA f = CFA franc; — = Not available.

## Table A2 Burkina Faso

| Season | Lint production (thousand tons) | Area (thousand hectares) | Yield (kg lint/hectare) | Ginning ratio (%) | Grower price (CFA f/kg seed cotton) | GDP deflator (2000 = 1.00) |
|---|---|---|---|---|---|---|
| 1970/71 | 8 | 81 | 105 | 35.9 | 32 | 0.20 |
| 1971/72 | 10 | 74 | 141 | 37.2 | 32 | 0.20 |
| 1972/73 | 12 | 70 | 171 | 36.7 | 32 | 0.22 |
| 1973/74 | 10 | 67 | 147 | 36.8 | 35 | 0.22 |
| 1974/75 | 11 | 62 | 184 | 37.1 | 40 | 0.25 |
| 1975/76 | 18 | 68 | 267 | 35.8 | 40 | 0.27 |
| 1976/77 | 20 | 79 | 255 | 36.6 | 40 | 0.28 |
| 1977/78 | 14 | 69 | 202 | 36.5 | 55 | 0.34 |
| 1978/79 | 22 | 72 | 312 | 37.3 | 55 | 0.39 |
| 1979/80 | 29 | 82 | 350 | 37.0 | 55 | 0.42 |
| 1980/81 | 23 | 75 | 311 | 37.3 | 55 | 0.45 |
| 1981/82 | 22 | 65 | 331 | 37.6 | 62 | 0.51 |
| 1982/83 | 29 | 72 | 400 | 38.1 | 62 | 0.56 |
| 1983/84 | 30 | 77 | 392 | 37.9 | 70 | 0.59 |
| 1984/85 | 34 | 82 | 418 | 39.0 | 90 | 0.63 |
| 1985/86 | 46 | 94 | 489 | 39.8 | 100 | 0.66 |
| 1986/87 | 66 | 127 | 520 | 39.8 | 100 | 0.61 |
| 1987/88 | 59 | 170 | 344 | 39.6 | 95 | 0.62 |
| 1988/89 | 59 | 171 | 344 | 40.3 | 95 | 0.64 |
| 1989/90 | 62 | 150 | 416 | 41.0 | 95 | 0.67 |
| 1990/91 | 77 | 166 | 465 | 40.8 | 95 | 0.68 |
| 1991/92 | 69 | 186 | 373 | 41.4 | 95 | 0.66 |
| 1992/93 | 69 | 177 | 392 | 42.4 | 85 | 0.66 |
| 1993/94 | 51 | 150 | 339 | 43.6 | 115 | 0.64 |
| 1994/95 | 63 | 184 | 341 | 43.9 | 115 | 0.76 |
| 1995/96 | 64 | 170 | 377 | 42.4 | 165 | 0.82 |
| 1996/97 | 90 | 196 | 460 | 42.1 | 180 | 0.88 |
| 1997/98 | 140 | 295 | 476 | 41.5 | 180 | 0.90 |
| 1998/99 | 119 | 355 | 335 | 41.8 | 185 | 0.97 |
| 1999/2000 | 109 | 245 | 445 | 42.9 | 185 | 0.95 |
| 2000/01 | 116 | 260 | 446 | 42.0 | 170 | 1.00 |
| 2001/02 | 158 | 359 | 440 | 41.8 | 200 | 1.05 |
| 2002/03 | 170 | 405 | 420 | 42.1 | 175 | 1.09 |
| 2003/04 | 204 | 459 | 444 | 42.2 | 185 | 1.11 |
| 2004/05 | 264 | 566 | 467 | 41.9 | 210 | 1.13 |
| 2005/06 | 298 | 646 | 462 | 41.9 | 175 | 1.16 |
| 2006/07 | 282 | 716 | 394 | 42.0 | 165 | — |

*Source:* SOFITEX for cotton statistics; IMF International Financial Statistics and World Bank estimates for GDP deflator.

*Note:* CFA f = CFA franc; — = Not available.

## Table A3 Cameroon

| Season | Lint production (thousand tons) | Area (thousand hectares) | Yield (kg lint/hectare) | Ginning ratio (%) | Grower price (CFA f/kg seed cotton) | GDP deflator (2000 = 1.00) |
|---|---|---|---|---|---|---|
| 1970/71 | 14 | 102 | 139 | 36.9 | 30 | 0.16 |
| 1971/72 | 16 | 99 | 160 | 36.6 | 31 | 0.17 |
| 1972/73 | 17 | 88 | 191 | 37.0 | 38 | 0.18 |
| 1973/74 | 10 | 61 | 170 | 37.3 | 40 | 0.19 |
| 1974/75 | 15 | 65 | 234 | 37.7 | 45 | 0.21 |
| 1975/76 | 19 | 73 | 261 | 38.5 | 45 | 0.23 |
| 1976/77 | 18 | 60 | 303 | 38.1 | 55 | 0.27 |
| 1977/78 | 15 | 48 | 317 | 37.8 | 65 | 0.29 |
| 1978/79 | 23 | 47 | 495 | 39.2 | 65 | 0.30 |
| 1979/80 | 31 | 57 | 544 | 38.4 | 70 | 0.34 |
| 1980/81 | 32 | 65 | 494 | 38.2 | 80 | 0.38 |
| 1981/82 | 31 | 63 | 486 | 38.5 | 90 | 0.42 |
| 1982/83 | 29 | 55 | 523 | 39.5 | 105 | 0.47 |
| 1983/84 | 37 | 71 | 519 | 39.0 | 117 | 0.53 |
| 1984/85 | 38 | 73 | 522 | 39.2 | 130 | 0.60 |
| 1985/86 | 46 | 89 | 514 | 39.7 | 140 | 0.67 |
| 1986/87 | 48 | 94 | 513 | 39.5 | 150 | 0.67 |
| 1987/88 | 45 | 95 | 476 | 39.6 | 140 | 0.66 |
| 1988/89 | 69 | 112 | 614 | 41.4 | 140 | 0.66 |
| 1989/90 | 43 | 89 | 482 | 41.3 | 95 | 0.65 |
| 1990/91 | 47 | 94 | 496 | 41.1 | 95 | 0.66 |
| 1991/92 | 47 | 90 | 524 | 41.2 | 95 | 0.68 |
| 1992/93 | 53 | 99 | 534 | 41.9 | 85 | 0.67 |
| 1993/94 | 52 | 103 | 503 | 40.9 | 130 | 0.69 |
| 1994/95 | 63 | 141 | 445 | 41.1 | 155 | 0.76 |
| 1995/96 | 79 | 159 | 495 | 40.3 | 180 | 0.89 |
| 1996/97 | 90 | 191 | 471 | 41.2 | 180 | 0.94 |
| 1997/98 | 73 | 172 | 425 | 40.2 | 190 | 0.97 |
| 1998/99 | 78 | 173 | 453 | 40.3 | 195 | 0.98 |
| 1999/2000 | 78 | 172 | 455 | 40.7 | 165 | 0.97 |
| 2000/01 | 96 | 199 | 482 | 41.6 | 225 | 1.00 |
| 2001/02 | 103 | 211 | 487 | 41.6 | 175 | 1.03 |
| 2002/03 | 95 | 181 | 526 | 41.1 | 180 | 1.07 |
| 2003/04 | 100 | 208 | 480 | 41.2 | 185 | 1.09 |
| 2004/05 | 125 | 215 | 581 | 40.8 | 190 | 1.09 |
| 2005/06 | 87 | 214 | 405 | 41.7 | 150 | 1.14 |
| 2006/07 | 77 | 202 | 382 | 42.0 | 175 | — |

*Source:* SODECOTON for cotton statistics; IMF International Financial Statistics and World Bank estimates for GDP deflator.
*Note:* CFA f = CFA franc; — = Not available.

139

**Table A4 Mali**

| Season | Lint production (thousand tons) | Area (thousand hectares) | Yield (kg lint/hectare) | Ginning ratio (%) | Grower price (CFA f/kg seed cotton) | GDP Deflator (2000 = 1.00) |
|---|---|---|---|---|---|---|
| 1970/71 | 20 | 66 | 303 | 37.7 | 25 | 0.13 |
| 1971/72 | 25 | 77 | 327 | 37.3 | 25 | 0.14 |
| 1972/73 | 24 | 77 | 315 | 36.8 | 25 | 0.15 |
| 1973/74 | 19 | 69 | 273 | 37.3 | 25 | 0.16 |
| 1974/75 | 23 | 69 | 332 | 37.6 | 38 | 0.17 |
| 1975/76 | 39 | 89 | 438 | 37.9 | 38 | 0.20 |
| 1976/77 | 45 | 110 | 412 | 38.3 | 38 | 0.23 |
| 1977/78 | 42 | 103 | 411 | 37.2 | 45 | 0.25 |
| 1978/79 | 48 | 118 | 407 | 37.7 | 45 | 0.27 |
| 1979/80 | 56 | 127 | 441 | 37.3 | 58 | 0.30 |
| 1980/81 | 41 | 111 | 367 | 37.6 | 58 | 0.35 |
| 1981/82 | 38 | 85 | 448 | 39.5 | 65 | 0.39 |
| 1982/83 | 50 | 105 | 474 | 39.1 | 65 | 0.44 |
| 1983/84 | 54 | 111 | 487 | 38.6 | 75 | 0.47 |
| 1984/85 | 55 | 119 | 464 | 38.4 | 75 | 0.53 |
| 1985/86 | 67 | 146 | 460 | 38.4 | 85 | 0.61 |
| 1986/87 | 79 | 152 | 518 | 39.0 | 85 | 0.55 |
| 1987/88 | 75 | 149 | 504 | 37.7 | 85 | 0.56 |
| 1988/89 | 97 | 190 | 511 | 39.0 | 85 | 0.55 |
| 1989/90 | 99 | 189 | 521 | 42.8 | 85 | 0.54 |
| 1990/91 | 115 | 205 | 558 | 41.5 | 93 | 0.57 |
| 1991/92 | 114 | 215 | 531 | 42.0 | 95 | 0.58 |
| 1992/93 | 135 | 246 | 547 | 42.2 | 95 | 0.59 |
| 1993/94 | 101 | 201 | 500 | 41.8 | 98 | 0.61 |
| 1994/95 | 128 | 270 | 475 | 43.7 | 130 | 0.78 |
| 1995/96 | 169 | 336 | 504 | 41.7 | 155 | 0.92 |
| 1996/97 | 190 | 420 | 451 | 41.9 | 155 | 0.97 |
| 1997/98 | 218 | 498 | 437 | 41.9 | 170 | 0.98 |
| 1998/99 | 217 | 504 | 431 | 41.6 | 185 | 0.98 |
| 1999/2000 | 197 | 482 | 409 | 42.8 | 150 | 0.95 |
| 2000/01 | 102 | 228 | 447 | 42.0 | 170 | 1.00 |
| 2001/02 | 240 | 532 | 451 | 42.0 | 200 | 1.00 |
| 2002/03 | 181 | 449 | 402 | 41.1 | 180 | 1.16 |
| 2003/04 | 254 | 549 | 464 | 41.0 | 200 | 1.19 |
| 2004/05 | 240 | 547 | 439 | 40.9 | 210 | 1.17 |
| 2005/06 | 222 | 551 | 403 | 41.5 | 168 | 1.20 |
| 2006/07 | 176 | 480 | 355 | 42.0 | 165 | — |

*Source:* CMDT for cotton statistics; IMF International Financial Statistics and World Bank estimates for GDP deflator; ICAC.

*Note:* CFA f = CFA franc; — = Not available.

## Table A5 Mozambique

| Season | Lint production (thousand tons) | Area (thousand hectares) | Yield (kg lint/hectare) | Ginning ratio (%) | Grower price (Mt/kg seed cotton) | GDP deflator (2000 = 100.00) |
|---|---|---|---|---|---|---|
| 1980/81 | 24 | 300 | 78 | — | — | 0.37 |
| 1981/82 | 18 | 249 | 74 | — | — | 0.38 |
| 1982/83 | 8 | 110 | 74 | — | — | 0.45 |
| 1983/84 | 7 | 100 | 74 | — | — | 0.51 |
| 1984/85 | 1 | 15 | 75 | — | — | 0.60 |
| 1985/86 | 4 | 48 | 77 | — | — | 0.79 |
| 1986/87 | 9 | 108 | 80 | — | — | 0.89 |
| 1987/88 | 7 | 131 | 51 | — | — | 2.52 |
| 1988/89 | 9 | 108 | 86 | — | — | 3.73 |
| 1989/90 | 8 | 65 | 123 | — | — | 5.50 |
| 1990/91 | 14 | 75 | 187 | 33 to 35 | 320 | 7.38 |
| 1991/92 | 12 | 68 | 169 | 33 to 35 | 479 | 11.88 |
| 1992/93 | 15 | 77 | 195 | 33 to 35 | 700 | 16.66 |
| 1993/94 | 16 | 77 | 205 | 33 to 35 | 1,100 | 24.73 |
| 1994/95 | 17 | 86 | 202 | 33 to 35 | 1,500 | 38.43 |
| 1995/96 | 16 | 244 | 67 | 33 to 35 | 3,900 | 54.21 |
| 1996/97 | 25 | 267 | 94 | 33 to 35 | 3,300 | 76.22 |
| 1997/98 | 31 | 327 | 95 | 33 to 35 | 2,950 | 83.86 |
| 1998/99 | 36 | 333 | 133 | 33 to 35 | 2,098 | 88.14 |
| 1999/2000 | 12 | 148 | 83 | 33 to 35 | 2,500 | 91.12 |
| 2000/01 | 24 | 234 | 104 | 35 to 38 | 2,700 | 100.00 |
| 2001/02 | 31 | 222 | 141 | 35 to 38 | 3,000 | 114.03 |
| 2002/03 | 19 | 200 | 111 | 35 to 38 | 3,800 | 134.21 |
| 2003/04 | 26 | 230 | 115 | 35 to 40 | 5,000 | 142.47 |
| 2004/05 | 26 | 230 | 115 | 35 to 40 | 5,000 | 165.98 |
| 2005/06 | 25 | 225 | 111 | 35 to 40 | 5,300 | 198.46 |
| 2006/07 | 43 | 233 | 185 | 36 | — | — |

*Source:* ICAC for cotton statistics; IMF International Financial Statistics and World Bank estimates for GDP deflator. Ginning ratios estimated based on interviews with Cotton Institute and ginners.

*Note:* Mt = Mozambican metical; — = Not available. Increase in production starting in 2000/01 related to entry of new firms.

## Table A6 Tanzania

| Season | Lint production (thousand tons) | Area (thousand hectares) | Yield (kg lint/hectare) | Ginning ratio (%) | Grower price (T Sh/kg seed cotton) | GDP deflator (2000 = 100.00) |
|---|---|---|---|---|---|---|
| 1980/81 | 43 | 350 | 123 | 32 | 3.7 | 1.61 |
| 1981/82 | 40 | 371 | 108 | 31 | 4.7 | 1.91 |
| 1982/83 | 44 | 446 | 99 | 31 | 6.0 | 2.36 |
| 1983/84 | 48 | 390 | 123 | 31 | 8.4 | 2.56 |
| 1984/85 | 31 | 370 | 84 | 29 | 13.0 | 2.96 |
| 1985/86 | 67 | 400 | 167 | 31 | 16.9 | 3.79 |
| 1986/87 | 78 | 450 | 173 | 31 | 19.5 | 4.94 |
| 1987/88 | 54 | 450 | 120 | 29 | 22.4 | 5.82 |
| 1988/89 | 35 | 260 | 135 | 31 | 28.0 | 10.33 |
| 1989/90 | 48 | 320 | 150 | 32 | 41.0 | 12.46 |
| 1990/91 | 85 | 450 | 189 | 32 | 70.0 | 15.26 |
| 1991/92 | 96 | 430 | 224 | 32 | 60.0 | 19.55 |
| 1992/93 | 45 | 344 | 131 | 31 | 80.0 | 24.51 |
| 1993/94 | 40 | 172 | 233 | 32 | 120.0 | 30.51 |
| 1994/95 | 82 | 344 | 238 | 33 | 207.0 | 40.01 |
| 1995/96 | 87 | 283 | 307 | 35 | 170.0 | 50.76 |
| 1996/97 | 62 | 350 | 177 | 30 | 185.0 | 60.56 |
| 1997/98 | 36 | 180 | 200 | 34 | 180.0 | 73.04 |
| 1998/99 | 35 | 250 | 142 | 35 | 175.0 | 83.42 |
| 1999/2000 | 41 | 182 | 227 | 33 | 180.0 | 93.03 |
| 2000/01 | 51 | 430 | 118 | 34 | 185.0 | 100.00 |
| 2001/02 | 63 | 392 | 161 | 33 | 182.0 | 107.17 |
| 2002/03 | 50 | 291 | 172 | 36 | 290.0 | 114.08 |
| 2003/04 | 118 | 459 | 257 | 35 | 250.0 | 120.49 |
| 2004/05 | 127 | 482 | 264 | 34 | 240.0 | 130.21 |
| 2005/06 | 72 | 434 | 165 | 35 | 365.0 | 137.16 |
| 2006/07 | 67 | 410 | 163 | 35 | — | — |

*Source:* ICAC for cotton statistics; IMF International Financial Statistics and World Bank estimates for GDP deflator.

*Note:* T Sh = Tanzania shilling; — = Not available.

## Table A7 Uganda

| Year | Lint production (tons) | Exchange rate (U sh/US$) | Consumer price index | World price of cotton | | Farmgate price for seed cotton | |
| | | | | Cotlook A Index for lint, nominal | | | |
| | | | | (U sh/kg) | (US$/kg) | Nominal (U sh/kg) | Real (U sh/kg) |
|---|---|---|---|---|---|---|---|
| 1990/91 | 8,000 | 481 | 0.32 | 895 | 1.86 | 340 | 1,052 |
| 1991/92 | 7,000 | 836 | 0.41 | 1,277 | 1.53 | 340 | 821 |
| 1992/93 | 9,000 | 1,190 | 0.63 | 1,538 | 1.29 | 200 | 317 |
| 1993/94 | 5,000 | 1,161 | 0.67 | 1,598 | 1.38 | 300 | 448 |
| 1994/95 | 6,105 | 941 | 0.73 | 1,789 | 1.90 | 400 | 545 |
| 1995/96 | 10,437 | 993 | 0.80 | 2,017 | 2.03 | 350 | 439 |
| 1996/97 | 20,480 | 1,049 | 0.86 | 1,835 | 1.75 | 320 | 374 |
| 1997/98 | 5,920 | 1,112 | 0.91 | 1,879 | 1.69 | 404 | 442 |
| 1998/99 | 15,170 | 1,296 | 0.91 | 1,776 | 1.37 | 313 | 343 |
| 1999/2000 | 21,645 | 1,497 | 0.97 | 1,718 | 1.15 | 253 | 260 |
| 2000/01 | 18,500 | 1,709 | 1.00 | 2,290 | 1.34 | 360 | 360 |
| 2001/02 | 22,200 | 1,749 | 1.02 | 1,687 | 0.96 | 262 | 257 |
| 2002/03 | 20,350 | 1,835 | 1.02 | 2,033 | 1.11 | 476 | 468 |
| 2003/04 | 29,600 | 1,967 | 1.10 | 2,915 | 1.48 | 615 | 561 |
| 2004/05 | 46,990 | 1,762 | 1.13 | 2,207 | 1.25 | 350 | 309 |
| 2005/06 | 18,981 | 1,806 | 1.23 | 2,262 | 1.25 | 400 | 325 |
| 2006/07 | 24,790 | 1,824 | 1.31 | 2,316 | 1.27 | 450 | 344 |
| 2007/08 | 27,750 | 1,710 | 1.37 | 2,445 | 1.43 | 450 | 329 |

*Sources:* CDO (production and farmgate prices); IMF (exchange rate and consumer price index); *Cotton Outlook* (Cotlook A Index); and authors' calculations.

*Note:* U sh = Uganda shilling. With the exception of consumer price index, all variables calculated on a crop year basis, April–March.

**Table A8 Zambia**

| Season | Lint production (thousand tons) | Area (thousand hectares) | Yield (kg lint/hectare) | Ginning ratio (%) | Grower price (K/kg seed cotton) | GDP deflator (2000 = 100.00) |
|---|---|---|---|---|---|---|
| 1980/81 | 6 | 38 | 159 | — | — | 0.04 |
| 1981/82 | 5 | 25 | 188 | — | — | 0.04 |
| 1982/83 | 12 | 34 | 337 | — | — | 0.04 |
| 1983/84 | 16 | 56 | 283 | — | — | 0.05 |
| 1984/85 | 11 | 55 | 199 | — | — | 0.06 |
| 1985/86 | 12 | 50 | 239 | — | — | 0.08 |
| 1986/87 | 7 | 38 | 190 | — | — | 0.15 |
| 1987/88 | 24 | 78 | 308 | — | — | 0.24 |
| 1988/89 | 12 | 91 | 133 | — | — | 0.32 |
| 1989/90 | 9 | 64 | 140 | — | — | 0.58 |
| 1990/91 | 20 | 92 | 219 | — | — | 1.20 |
| 1991/92 | 9 | 62 | 140 | — | — | 2.32 |
| 1992/93 | 12 | 68 | 169 | — | — | 6.16 |
| 1993/94 | 13 | 74 | 170 | — | — | 15.00 |
| 1994/95 | 17 | 65 | 262 | — | 521 | 24.81 |
| 1995/96 | 20 | 115 | 174 | 38 | 558 | 34.25 |
| 1996/97 | 35 | 140 | 250 | 38 | 534 | 42.10 |
| 1997/98 | 42 | 173 | 243 | 38 | 570 | 53.03 |
| 1998/99 | 36 | 150 | 240 | 39 | 444 | 63.37 |
| 1999/2000 | 30 | 150 | 200 | 39 | 680 | 76.90 |
| 2000/01 | 30 | 114 | 263 | 39 | 840 | 100.00 |
| 2001/02 | 46 | 165 | 279 | 40 | 860 | 128.03 |
| 2002/03 | 47 | 150 | 313 | 40 | 1,220 | 159.19 |
| 2003/04 | 69 | 254 | 271 | 40 | 1,420 | 192.54 |
| 2004/05 | 81 | 275 | 295 | 41 | 1,220 | 230.46 |
| 2005/06 | 79 | 275 | 286 | 41 | 850 | 270.61 |
| 2006/07 | 35 | 180 | 194 | 42 | 850 | — |

*Source:* ICAC for cotton statistics; IMF International Financial Statistics and World Bank estimates for GDP deflator. Ginning ratio estimated from interviews with ginners and Cotton Development Trust (CDT).

*Note:* K = Zambian kwacha; — = Not available.

## Table A9 Zimbabwe

| Season | Lint production (thousand tons) | Area (thousand hectares) | Yield (kg lint/hectare) | Ginning ratio (%) | Grower price (Z$/kg seed cotton) | GDP deflator (2000 = 100.00) |
|---|---|---|---|---|---|---|
| 1980/81 | 62 | 134 | 459 | 36 | 0.40 | 2.21 |
| 1981/82 | 56 | 112 | 497 | 42 | 0.52 | 2.53 |
| 1982/83 | 60 | 138 | 435 | 41 | 0.52 | 2.89 |
| 1983/84 | 91 | 190 | 481 | 41 | 0.57 | 3.45 |
| 1984/85 | 103 | 231 | 447 | 38 | 0.67 | 3.57 |
| 1985/86 | 89 | 192 | 462 | 35 | 0.75 | 3.80 |
| 1986/87 | 87 | 243 | 356 | 31 | 0.80 | 4.24 |
| 1987/88 | 116 | 272 | 427 | 34 | 0.85 | 4.53 |
| 1988/89 | 92 | 248 | 371 | 34 | 1.11 | 5.31 |
| 1989/90 | 67 | 228 | 293 | 33 | 1.35 | 6.26 |
| 1990/91 | 72 | 273 | 262 | 28 | 1.63 | 7.19 |
| 1991/92 | 21 | 235 | 88 | 35 | 1.35 | 9.39 |
| 1992/93 | 75 | 246 | 304 | 35 | 2.62 | 11.98 |
| 1993/94 | 60 | 230 | 261 | 33 | 3.20 | 14.64 |
| 1994/95 | 38 | 194 | 194 | 38 | 3.70 | 17.72 |
| 1995/96 | 104 | 264 | 394 | 37 | 4.20 | 19.41 |
| 1996/97 | 101 | 313 | 322 | 36 | 6.00 | 24.42 |
| 1997/98 | 105 | 286 | 368 | 38 | 9.00 | 28.37 |
| 1998/99 | 115 | 330 | 349 | 38 | 15.00 | 38.80 |
| 1999/2000 | 138 | 369 | 374 | 39 | 18.00 | 64.01 |
| 2000/01 | 135 | 389 | 347 | 40 | 28.00 | 100.00 |
| 2001/02 | 80 | 363 | 221 | 41 | 57.00 | 176.57 |
| 2002/03 | 103 | 327 | 315 | 42 | 400.00 | 394.20 |
| 2003/04 | 130 | 330 | 395 | 39 | 1,800.00 | 2,014.74 |
| 2004/05 | 76 | 320 | 237 | 38 | 4,500.00 | 9,063.71 |
| 2005/06 | 115 | 380 | 303 | 44 | 80,000.00 | 28,208.97 |
| 2006/07 | 104 | 400 | 261 | 41 | — | — |

*Source:* ICAC for cotton statistics; IMF International Financial Statistics and World Bank estimates for GDP deflator.

*Note:* Z$ = Zimbabwe dollar; — = Not available.

**Table A10 Ginning and FOB-to-CIF Costs, All WCA Countries, 1970 to 2006**
*(nominal terms)*

| Season | Nominal Cotlook A Index ($/kg) | Exchange rate (CFA f/$) | Ginning costs (CFA f/kg) | Sea freight costs ($/ton) | Marketing costs ($/ton) | FOB-to-CIF costs CFA f/kg of lint | FOB-to-CIF costs % of the Cotlook A Index |
|---|---|---|---|---|---|---|---|
| 1970/71 | 0.69 | 276 | 50 | 35 | 21 | 15 | 8.1 |
| 1971/72 | 0.82 | 262 | 50 | 35 | 25 | 15 | 7.3 |
| 1972/73 | 0.92 | 237 | 50 | 40 | 28 | 15 | 7.4 |
| 1973/74 | 1.69 | 233 | 55 | 50 | 51 | 24 | 6.0 |
| 1974/75 | 1.16 | 220 | 60 | 55 | 35 | 19 | 7.8 |
| 1975/76 | 1.44 | 228 | 65 | 60 | 43 | 24 | 7.2 |
| 1976/77 | 1.84 | 248 | 75 | 60 | 55 | 28 | 6.3 |
| 1977/78 | 1.43 | 237 | 80 | 65 | 43 | 25 | 7.5 |
| 1978/79 | 1.68 | 216 | 90 | 70 | 50 | 26 | 7.2 |
| 1979/80 | 1.88 | 208 | 100 | 75 | 56 | 28 | 7.0 |
| 1980/81 | 2.08 | 243 | 110 | 80 | 62 | 37 | 6.9 |
| 1981/82 | 1.63 | 301 | 120 | 85 | 49 | 43 | 8.2 |
| 1982/83 | 1.69 | 359 | 130 | 90 | 51 | 52 | 8.3 |
| 1983/84 | 1.93 | 414 | 140 | 110 | 58 | 72 | 8.7 |
| 1984/85 | 1.52 | 472 | 145 | 120 | 46 | 75 | 10.9 |
| 1985/86 | 1.08 | 378 | 150 | 110 | 32 | 50 | 13.2 |
| 1986/87 | 1.37 | 316 | 150 | 100 | 41 | 43 | 10.3 |
| 1987/88 | 1.60 | 295 | 135 | 95 | 48 | 42 | 8.9 |
| 1988/89 | 1.46 | 308 | 120 | 95 | 44 | 44 | 9.5 |
| 1989/90 | 1.77 | 305 | 125 | 95 | 53 | 42 | 8.4 |
| 1990/91 | 1.81 | 273 | 120 | 90 | 54 | 40 | 8.0 |
| 1991/92 | 1.50 | 281 | 115 | 90 | 45 | 36 | 9.0 |
| 1992/93 | 1.30 | 269 | 120 | 90 | 39 | 36 | 9.9 |
| 1993/94 | 1.50 | 411 | 150 | 90 | 45 | 68 | 9.0 |
| 1994/95 | 2.02 | 522 | 175 | 90 | 61 | 77 | 7.5 |
| 1995/96 | 1.91 | 500 | 190 | 90 | 57 | 75 | 7.7 |
| 1996/97 | 1.76 | 538 | 200 | 80 | 53 | 75 | 7.5 |
| 1997/98 | 1.64 | 598 | 210 | 75 | 49 | 74 | 7.6 |
| 1998/99 | 1.35 | 593 | 225 | 70 | 40 | 67 | 8.2 |
| 1999/2000 | 1.20 | 649 | 230 | 65 | 36 | 70 | 8.4 |
| 2000/01 | 1.27 | 731 | 245 | 60 | 38 | 72 | 7.7 |
| 2001/02 | 0.97 | 731 | 225 | 55 | 29 | 60 | 8.6 |
| 2002/03 | 1.16 | 648 | 220 | 60 | 35 | 57 | 8.2 |
| 2003/04 | 1.47 | 555 | 215 | 55 | 38 | 50 | 6.3 |
| 2004/05 | 1.23 | 524 | 220 | 55 | 32 | 46 | 7.1 |
| 2005/06 | 1.24 | 535 | 225 | 60 | 32 | 49 | 7.4 |

*Sources:* Country sources (see tables A1 through A9), World Bank, and IMF International Financial Statistics.

*Note:* FOB = Free on board; CIF = Cost, insurance, and freight. According to ICAC classification, "All WCA" includes the four WCA countries in this study (Benin, Burkina Faso, Cameroon, and Mali) plus the Central African Republic, Chad, Côte d'Ivoire, Guinea, Niger, Madagascar, Senegal, and Togo. The Cotlook A Index (cotton lint) and the CFA f/$ exchange rate were calculated as averages over March through July to account for the fact that most cotton is marketed during this period. Marketing costs were calculated as 3 percent of the Cotlook A Index through 2002/03 and 2.6 percent afterward. The ginning costs are as reported by the cotton companies (averages for the entire WCA) and are expressed in terms of cotton lint.

## Table A11 Cotton Production, Area, and Yields, World and All WCA Countries, 1970–2006

| Season | World | | | All WCA | | |
|---|---|---|---|---|---|---|
| | Production (thousand tons) | Area (thousand hectares) | Yield (kgs/hectare) | Production (thousand tons) | Area (thousand hectares) | Yield (kgs/hectare) |
| 1970/71 | 11,740 | 31,778 | 369 | 109 | 644 | 169 |
| 1971/72 | 12,938 | 33,024 | 392 | 141 | 686 | 205 |
| 1972/73 | 13,595 | 33,818 | 402 | 143 | 643 | 222 |
| 1973/74 | 13,615 | 32,558 | 418 | 138 | 616 | 224 |
| 1974/75 | 13,926 | 33,285 | 418 | 158 | 629 | 252 |
| 1975/76 | 11,706 | 30,001 | 390 | 190 | 723 | 263 |
| 1976/77 | 12,385 | 31,513 | 393 | 195 | 712 | 274 |
| 1977/78 | 13,860 | 34,966 | 396 | 179 | 666 | 268 |
| 1978/79 | 12,933 | 34,000 | 380 | 216 | 702 | 308 |
| 1979/80 | 14,084 | 33,100 | 425 | 235 | 652 | 360 |
| 1980/81 | 13,831 | 33,667 | 411 | 204 | 633 | 323 |
| 1981/82 | 14,991 | 33,948 | 442 | 202 | 551 | 366 |
| 1982/83 | 14,479 | 32,569 | 445 | 253 | 591 | 427 |
| 1983/84 | 14,499 | 32,137 | 451 | 279 | 675 | 413 |
| 1984/85 | 19,247 | 35,217 | 547 | 327 | 708 | 461 |
| 1985/86 | 17,461 | 32,792 | 532 | 351 | 838 | 419 |
| 1986/87 | 15,269 | 29,503 | 518 | 409 | 846 | 483 |
| 1987/88 | 17,609 | 31,238 | 564 | 414 | 911 | 454 |
| 1988/89 | 18,301 | 33,522 | 546 | 498 | 1,101 | 452 |
| 1989/90 | 17,365 | 31,640 | 549 | 458 | 1,026 | 446 |
| 1990/91 | 18,978 | 33,050 | 574 | 533 | 1,118 | 477 |
| 1991/92 | 20,677 | 34,710 | 596 | 521 | 1,232 | 423 |
| 1992/93 | 17,943 | 32,238 | 557 | 539 | 1,209 | 446 |
| 1993/94 | 16,861 | 30,430 | 554 | 510 | 1,176 | 434 |
| 1994/95 | 18,762 | 32,114 | 584 | 573 | 1,398 | 410 |
| 1995/96 | 20,330 | 36,056 | 564 | 667 | 1,502 | 444 |
| 1996/97 | 19,599 | 34,111 | 575 | 790 | 1,753 | 451 |
| 1997/98 | 20,094 | 33,746 | 595 | 921 | 2,120 | 435 |
| 1998/99 | 18,705 | 32,846 | 569 | 858 | 2,202 | 390 |
| 1999/00 | 19,095 | 31,929 | 598 | 851 | 2,034 | 418 |
| 2000/01 | 19,457 | 31,766 | 612 | 693 | 1,668 | 415 |
| 2001/02 | 21,500 | 33,396 | 644 | 994 | 2,258 | 440 |
| 2002/03 | 19,297 | 29,872 | 646 | 913 | 2,128 | 429 |
| 2003/04 | 20,714 | 32,021 | 647 | 906 | 2,216 | 409 |
| 2004/05 | 26,290 | 35,332 | 744 | 1,119 | 2,474 | 452 |
| 2005/06 | 24,752 | 34,252 | 723 | 923 | 2,349 | 393 |

*Source:* World data are from ICAC; WCA data are from the cotton companies.

*Note:* According to ICAC classification, "All WCA" includes the four WCA countries in this study (Benin, Burkina Faso, Cameroon, and Mali) plus the Central African Republic, Chad, Côte d'Ivoire, Guinea, Niger, Madagascar, Senegal, and Togo.

# Notes

1. FAOSTAT for total agricultural trade, ICAC for cotton trade.

2. See chapter 2 on cotton's market context for more detail.

3. While some might argue that Tanzania is an exception, the persistent and very serious efforts by the government and private stakeholders to resolve the input supply problem in the sector suggest that external input is considered critical.

4. See Glover (1990) for a review of experience in eastern and southern Africa through the late 1980s.

5. These low costs of production are related primarily to the very low price at which many smallholder farmers are willing to "sell" their labor in production of the crop, and to the low supervisory costs inherent in using primarily family labor. See Binswanger and McIntire (1987).

6. See Jaffee (1994), however, for an empirical review of the widely varying circumstances under which contract farming has emerged, and examples of failure where external conditions seemed favorable.

7. One study did look at cross-country experience in selected countries of WCA and ESA and considered the pros and cons of different institutional structures (Goreux and Macrae 2002). Compared with that study, the present work is more comprehensive, has substantially more coverage of ESA, and has the benefit of an additional five years of post-reform experience in the study countries.

8. It is worth noting, however, that production in the two major WCA producers has dropped sharply in 2007/08 compared with 2006/07: from 700,000 to 360,000 metric tons (a drop of 48 percent) in Burkina Faso and from 442,000 to 243,000 metric tons (a drop of 45 percent) in Mali.

9. Except Benin.

10. Mozambique is a special case to be discussed later.

11. Important cotton sector reforms have taken place in Côte d'Ivoire in the past 10 years, but it is not part of the study sample.

12. Although India's area allocated to GM cotton (10 percent) is small compared with other countries, its share in worldwide GM cotton production is high because the total cotton area in India is high and GM yields are well above mean yields in the country.

13. Price variability of cotton has not been that different from other primary commodities. Pan and Valderrama (2005), for example, compared the price variability of 22 primary commodities and concluded that during 2000–04 17 commodities exhibited more price variability than cotton. Similarly, Gilbert (2006) ranked 21 commodities according to their volatility and found cotton to be somewhere in the middle.

14. A study that estimated the price transmission elasticities from crude oil to 35 primary commodities (including cotton and most food commodities) found that the average elasticity for

food commodities was 0.18 and highly significant while that for cotton was 0.14 and marginally significant (Baffes 2007). Although this result seems counterintuitive because cotton competes with man-made fibers, whose key input is crude oil, food commodities are much bulkier than cotton while the crude oil component in chemical fibers is small. Similarly, the study found low transmission elasticities for natural rubber (which competes with synthetic rubber) as well as some energy-intensive metals.

15. For a discussion of how the US dollar exchange rate affects dollar-commodity prices, see Radetzki (1985).

16. The highly divergent results of these models reflect a number of factors. First, there are differences in the level and structure of support. For example, some models incorporate China's support to its cotton sector and model its removal; others do not. Second, there are differences in the underlying scenarios. Some models assume liberalization in all commodity markets while others assume liberalization only in the cotton sector. Third, the models use different base years and hence different levels of subsidies. For example, support in the United States was three times as high in 1999 as in 1997.

17. Note that these differentials are between short staple varieties and extra long staple pima varieties grown in different environments. As will be shown in chapter 7, the price differentials observed within African upland cotton varieties, while still important, are much smaller than this.

18. Fiber length is the average length of the longest half of fibers. Grade is a commercial value based on a visual assessment of a combination of lint color, cleanliness, and preparation. Color is determined by the degree of reflectance (good) and yellowness (bad). Micronaire is a measure of fiber fineness and maturity.

19. Neps are cotton fibers tangled into a knot.

20. *Gossypium barbadense*.

21. Cost and Freight Far East.

22. Ring spun carded yarn is typically used for knitting and weaving, in a large range of coarse to fine counts.

23. Combed yarns are stronger, more uniform, smoother, purer, and have greater shine than carded yarns.

24. In Africa, these mother companies include Dunavant, Cargill, Plexus, Dagris, Reinhart, and others.

25. Or the quotation for the African franc zone in *Cotton Outlook*.

26. Though exceptions may exist: for example, supply is more atomized in Benin than in Zambia.

27. In some countries, seeds are sold directly on the domestic market without processing into oil and cake. In Mali, about 20 percent of cotton seeds are sold as livestock feed at prices 30 percent higher than those paid by the oil mills.

28. In most cases this processing takes place within the country of origin. However, varying quantities of seed are exported from Mozambique and Zambia to South Africa for

processing. Similarly, at the end of the 1990s, there was a European demand for seeds. Some WCA cotton companies decided then to export, bringing financial problems to the large-scale oil industry in the countries concerned, which needs large quantities of seeds to cover fixed costs. This export market to Europe has since declined and only one trader from WCA is still dealing on it.

29. Because the world market for cottonseed oil is very thin, its price indicator may not be as reliable as those of the four major oils.

30. CFDT was renamed DAGRIS in 2000.

31. CFDT withdrew from Benin after the advent of the socialist regime in that country in 1972.

32. With the exception of Benin and Burkina Faso, where cotton extension was provided by the national agricultural extension systems.

33. Predefined standard costs on the basis of which cotton companies were remunerated for the public services they provided.

34. Benin and Côte d'Ivoire were exceptions to this single channel pattern. Reforms in Benin started earlier than elsewhere but were complex and badly managed. In Côte d'Ivoire, full liberalization took place, but was soon affected by the country's political crisis.

35. Côte d'Ivoire and Senegal in WCA have also introduced private cotton companies.

36. Eventually adjusted before harvest in case of major changes in world prices.

37. Distributors are farmers chosen from the community to help mainly in the logistics of credit provision given the large number of farmers that Dunavant deals with. Efforts at more serious extension assistance have been carried out through Dunavant's YIELD program, financed by the German development agency GTZ.

38. Input costs are subsidized by companies, rather than provided on credit. Because the subsidy must eventually be recovered in the price, it can be conceived of as partial in-kind credit.

39. See Leibenstein (1966) for original concept of "x-inefficiency," and vast follow-up literature in the area of business management.

40. The details of this system are explained in more detail in chapter 6.

41. By labeling soil fertility "exogenous," the intention is not to dismiss concerns about the impact (negative or positive) of cotton production on soil fertility. Rather, it points out that other factors also make a major contribution to observed soil fertility, the inherent characteristics of the soil, population density, and the overall management of the farming system being three major ones.

42. Consistent with this definition, this study does not cover the further processing of lint because it belongs to a completely different, downstream segment of the cotton and textile value chain.

43. *Filière* is the French word for value chain or subsector and embodies the vertically coordinated approach long pursued in WCA.

44. In 2004, the government of Zimbabwe did get involved in resolving a pricing dispute between producers and companies.

45. And in Cameroon, the price support fund has been exhausted by the high prices.

46. See chapter 1 of this book for background on this issue.

47. Benin is not included in this review because the complexity and uniqueness of its system does not lend itself to the general lessons we seek in this review.

48. Some of Burkina Faso's reported production is due to seed cotton coming over the border from Côte d'Ivoire and other countries, spurred by the unrest in Côte d'Ivoire and high prices paid by SOFITEX.

49. Though Uganda has been classified as a hybrid sector, it is included in the discussion here with Tanzania under competitive sectors because its competitive structure since reform has created problems very similar to those found in Tanzania and has driven the types of hybrid institutional set-ups described earlier.

50. Seed has so far been included in the system but there has been debate as to whether it should be removed in the future.

51. For more detail, see Tschirley, Poulton, and Boughton (2008) and Poulton and Hanyani-Mlambo (2007).

52. Cargill purchased Clark in 2006.

53. A sharp appreciation of the Zambian kwacha from late 2006 through May or June 2007 was also a major factor.

54. See concept of loss aversion in behavioral economics, in which "the disutility of giving up an object is greater than the utility associated with acquiring it" (Kahneman, Knetsch, and Thaler 1991, 194).

55. In the international market, cotton is priced in US cents per pound. One cent per pound is equal to 2.2046 cents per kg.

56. Growing fewer varieties in a country makes it easier to maintain homogeneity of quality, though proper controls (as in Zambia) and good classification can ensure good performance even when several varieties are grown.

57. Assuming that the whole crop was sold on a given day.

58. Actual season-average contract prices can be different from these theoretical averages depending on the timing of sales. In addition, final average realized prices can differ from contract prices depending on the actual quality and weight shipped and on eventual claims. Nevertheless, these calculations represent the current best estimates of average market quality premiums earned by each country's sector.

59. It is common for five or more buyers, all with their own buying posts, to be in a single village. Companies either hire enterprising local residents to run buying posts at harvest time or contract with the staff of the local primary cooperative society to buy on their behalf. Finding trustworthy agents is sometimes a challenge in itself, without adding quality control to their responsibilities.

60. International firms feature more prominently in the Uganda sector than in Tanzania, but there are still plenty of local firms in the market.

61. Some farmers, principally those delivering to smaller companies, also began to engage in the kind of opportunistic practices seen in Tanzania. Cottco and Cargill's buying operations allow them to trace seed cotton back to the farmers who sold it to them, so they have been less affected by these practices. However, even they report a dramatic increase in foreign matter found within seed cotton bales and have had to employ additional staff to sort through all bales before they are sent to the gins.

62. For some years the NCC was chaired by a representative of one of the oil processors, ensuring that sector policy protected their interests, as well as those of ginners and producers.

63. Demand for cake in Zimbabwe would certainly have been strong in the 1990s, when a strong commercial livestock industry existed.

64. Agricultural research in Africa is overwhelmingly funded by the public sector and at much lower levels than in developed countries relative to the agricultural sector's contribution to GDP. Growth in agricultural funding in Africa slowed dramatically in the 1990s compared with previous decades, even as numbers of researchers increased. For a thorough review of agricultural research in Africa, see Beintema and Stads (2006).

65. In virtually all countries, research plays an important role in basic seed multiplication, as well as in development of new varieties.

66. Similarly, it is desirable for producers' associations to be involved in setting research priorities and monitoring research performance.

67. Within a competitive system, a high degree of free-riding on such effort might be expected.

68. Note that Lele, Van de Walle, and Gbetibouo (1989) included the potential collapse of research systems in their warnings about dissolving the single-channel cotton systems throughout WCA if viable alternative institutional setups were not in place.

69. It also complicates the priority setting for research, increasing the chances of conflicts of interest between producers, ginners, and lint buyers.

70. Field surveys were not carried out in Benin; thus, Benin does not appear in the comparative analysis in this chapter.

71. Yields in Benin, Burkina Faso, Chad, and Mali, have all followed this general pattern.

72. This calculation excludes the labor figures provided by focus group respondents in Mozambique, which were very high. It is not clear whether the very high labor figures in Mozambique were a result of different data collection methods used there or reflected actual differences in labor use. It is true that high prevalence of malaria and malnutrition (especially during 2005/06) are believed to have depressed labor productivity in Mozambique.

73. In Mozambique, pre-emergence herbicide use is in a pilot phase in Cabo Delgado province. Farmers were very positive about this initiative, despite the increased costs, because it enabled them to weed their food crops on time. Cargill also provides herbicide to some farmers in low altitude areas of Zambia, where weed pressure is intense.

74. Group 1 in Zambia does use a foliar feed fertilizer that they receive on credit; this provides much less nitrogen than the fertilizers used in WCA, Zimbabwe, and Uganda, and is much less costly.

75. These percentage figures for Mozambique, Uganda, and Zambia are probably on the high side because of the high estimated costs for labor or hired services included in the group budgets. However, the basic point that a large proportion of producers in these countries achieve low returns to their labor input into cotton production is a robust one.

76. In Tanzania, each of the four villages contained 500 to 1,000 or more households, so it would have been too time-consuming to use the full list. In addition, while village leaders could always give a figure for the number of households in their village (based on the previous census and subsequent adjustments), complete lists of all household heads were not always available. In such cases, names were drawn at random from the lists for individual "wards" that were available.

77. The entire exercise took three or more hours per village. Researchers thus provided food part way through the process.

78. Several ginners were interviewed in both Tanzania and Zimbabwe and a composite picture of ginning costs was built up from the information on different cost components provided by different respondents. In Mozambique, only one ginner, considered to be representative of the least efficient companies, agreed to provide cost information.

79. In Zimbabwe and Zambia, a number of new roller gins are being installed. However, the majority of ginning capacity in these countries is still made up of saw gins.

80. This figure reflects uneconomic pricing (based on official exchange rates) by a state-owned enterprise within a highly distorted economy.

81. This claim is based on the higher level of research in WCA countries than in Mozambique and the history of substantial investment by WCA cotton companies in both infrastructure and research. Mozambican companies did have substantial road maintenance costs during the first several years of the privatization phase.

82. One study did look at cross-country experience in WCA (Benin, Burkina Faso, and Cote d'Ivoire; also Ghana) and ESA (Zimbabwe and, with less detail, Tanzania) and explicitly considered the pros and cons of different institutional structures (Goreux and Macrae 2002). The report spotlighted some findings similar to those in this present book (for example, the collapse of input supply and quality control in Tanzania). Compared with that study, the present work is more comprehensive, has substantially more coverage of ESA, and has the benefit of an additional five years of post-reform experience in the study countries.

83. By "path dependency," we mean that current structure is heavily influenced by past structure—the past structural path that the country has traveled.

84. Unfortunately, this calculation could only be performed for one year, which happened to be a bumper harvest year in Tanzania. It would be useful to extend these calculations to additional years, as has been done, for example, for seed cotton pricing and quality premiums.

85. As shown in chapter 7, in Zambia the efforts of the two dominant ginning companies resulted in an increase in the premium for the top type of Zambian lint relative to the Cotlook A Index by US$0.05 per lb over a period of five years. The two companies used different approaches, but with equal success in largely eliminating contamination: Dunavant relies primarily on the manual removal of contamination at the gin, while Cargill has succeeded in

154

changing farmer behavior to avoid contamination. In both cases, the limited alternative outlets through which farmers can sell their seed cotton means that they have to respond to the quality initiatives of the dominant companies.

86. As noted in this chapter, the competitive structure of the cotton sector in Tanzania may prevent it from ever achieving a high quality reputation. Instead, its combination of competitive structure and abundant area for production expansion makes it better placed to compete primarily on the basis of cost. Nevertheless, even as a low-cost producer Tanzania has to take urgent action to reduce the contamination of its lint. Its challenge is to find institutional arrangements appropriate to a competitive sector that can enhance the quality incentives facing both producers and ginners. A proposed village auction system for the purchase of seed cotton may be one such arrangement (Poulton and Maro 2007).

87. Cottco markets some of its lint directly to spinners, bypassing international merchants altogether. However, its lint marketing activities have been badly hit by both the national economic crisis (leading to cash shortages) and the changes in sector structure, which make it less able to predict what quantities of lint of a particular quality it is going to be able to produce. Hence, there has been a shift away from forward and cost, insurance, and freight sales toward spot market sales on a free-on-board or free-on-truck basis.

88. According to figure 10.1, there has been little improvement across ESA as a whole since liberalization, although individual countries, such as Zambia, have posted noteworthy improvements.

89. It also shows how the efficiency of technical assistance can be increased when it is channeled through a strong private sector partner.

90. A secure and well-regulated local monopoly arrangement may be ideal for this purpose because concessionaires can capture the immediate benefits of better pest management or demonstrate how their farmers have benefited from lower input costs, while eventual re-tendering of concessions should provide an incentive to seek such outcomes. This incentive may be weaker in a national monopoly system, unless there is pressure from farmer organizations, civil society groups (concerned about human health or environmental impacts), or politicians.

91. A second technology likely to be of interest to many farmers is low-volume herbicides. These are labor saving, so the poverty impact of their introduction can be questioned. However, as well as appealing to "larger" smallholder cotton producers, who currently rely heavily on hired labor (ESA) or animal traction with family labor (WCA) for weeding work, they might ultimately also assist poorer producers who struggle to allocate their labor across hiring out, food crop production, and cotton, with the latter currently suffering from untimely and inadequate labor input.

92. The experience of Bt cotton production by South African smallholders is instructive, in both the yield impact and rapid adoption of Bt (Thirtle et al. 2003) and in the dependence of that success on institutional arrangements to support provision of the relevant input on credit (Gouse 2007).

93. The main emphasis in this book has been on sector types and their influence on sector performance. However, the issue of firm types is clearly linked to that of sector types. Arguably, for example, concentrated and local monopoly sectors can perform well in part

because they allow a leading role for large companies, many of them affiliated with international cotton merchants. Conversely, the experiences of Zimbabwe since 2003 and Uganda since 2000 show that the presence of such firms within a sector is not a sufficient condition for strong performance, hence the chosen focus of this book.

94. Dominant firms have incentives to cooperate with each other in concentrated sectors, whereas in local monopolies the onus is on the regulatory body to monitor concessionaire companies that may be reluctant to disclose information. The costs of (re-)tendering concessions may also be higher than the costs of issuing licenses under a concentrated system and the (re-)tendering process may encourage firms to invest more in maintaining relations with the regulator than in improving performance in their monopoly areas.

95. In Côte d'Ivoire, the cotton sector has moved fairly rapidly from a local monopoly system toward a concentrated one, even though this evolution was not planned in advance and is partly due to the socio-political events that have taken place since the early 2000s.

96. Chapter 6 suggests reasons for these different approaches in the two countries.

97. Savings and Credit Cooperatives (SACCO).

# Bibliography

Badiane, Ousmane, Dhaneshwar Ghura, Louis Goreux, and Paul Masson. 2002. "Cotton Sector Strategies in West And Central Africa." Policy Research Working Paper No. 2867, World Bank, Washington, DC.

Baffes, J. 2005. "The Cotton Problem." *The World Bank Research Observer* 20 (1): 109–44.

———. 2007. "Oil Spills on Other Commodities." *Resources Policy* 32 (3): 126–34.

———. 2008. "Uganda Country Study." Background paper for Comparative Analysis of Organization and Performance of African Cotton Sectors: Learning from Reform Experience. World Bank, Washington, DC.

Baffes, J., Ibrahim Elbadawi, and Stephen O'Connell. 1999. "Single-Equation Estimation of the Equilibrium Real Exchange Rate." In *Exchange Rate Misalignment: Concepts and Measurements for Developing Countries*, ed. L. E. Hinkle and P. J. Montiel, 405–65. New York: Oxford University Press.

Beintema, Nienke M., and Gert-Jan Stads. 2006. "Agricultural R&D in Sub-Saharan Africa: An Era of Stagnation." Agricultural Science and Technology Indicators (ASTI) Initiative Background Report, IFPRI, Washington, DC.

Benfica, Rui, David Tschirley, and Liria Sambo. 2002. "The Impact of Alternative Agro-Industrial Investments on Poverty Reduction in Rural Mozambique." Research Report No. 51E, Ministry of Agriculture of Mozambique, Maputo. http://www.aec.msu.edu/agecon/fs2/mozambique/wps51e.pdf.

Binswanger, Hans, and J. McIntire. 1987. "Behavioral and Material Determinants of Production Relations in Land Abundant Tropical Agriculture." *Economic Development and Cultural Change* 36 (1): 73–99.

Boughton, D. 2008. "Mozambique Country Study." Background paper for Comparative Analysis of Organization and Performance of African Cotton Sectors: Learning from Reform Experience. World Bank, Washington, DC.

Chaudhry, Rafiq. 2007. "Outlook for Cotton Production and Developments in Production Research." Paper presented at the UNCTAD/UNDP Workshop on Enhancing the Cotton Value Chain in Africa through Trade and Investment, with a Special Emphasis on Regional and South-South Cooperation. Bamako, Mali, December 11–12.

Conseil Ouest et Centre Africain Pour La Recherche et le Développement Agricole (CORAF/WECARD). Undated. Evaluation des Unités Opérationelles du CORAF/WECARD. Rapport Final. Unpublished.

*Cotton Outlook*. 2005. Cotlook's GM Cotton Survey. *Special Issue—The ICAC 64th Plenary Meeting*. Liverpool, United Kingdom.

Delgado, Christopher. 1999. "Sources of Growth in Smallholder Agriculture in Sub-Saharan Africa: The Role of Vertical Integration of Smallholders with Processors and Marketers of High Value-Added Items." *Agrekon* 38: 165–89.

Devarajan, S. 1999. "Estimates of Real Exchange Rate Misalignment with a Simple General-Equilibrium Model." In *Exchange Rate Misalignment: Concepts and Measurements for Developing Countries*, ed. L. E. Hinkle and P. J. Montiel, 359–80. New York: Oxford University Press.

Estur, G. 2008. "Quality and Marketing of African Cotton." Background paper for Comparative Analysis of Organization and Performance of African Cotton Sectors: Learning from Reform Experience. World Bank, Washington, DC.

European Commission. 2003. "Agricultural Reform Continued: Commission Proposes Sustainable Agricultural Model for Europe's Tobacco, Olive Oil and Cotton Sectors." Press Release, European Commission, September 23, Brussels.

FAO Statistics. Various reports.

FAO (Food and Agriculture Organization of the United Nations). 2004. "Cotton: Impact of Support Policies in Developing Countries." Trade Policy Technical Note No. 1. ftp://ftp.fao.org/docrep/fao/007/y5533e/y5533e01.pdf.

Gergely, N. 2004. "Analyse comparée des coûts de la filière coton au Burkina, au Mali et au Cameroun." (Cost analysis of the cotton supply chain in Mali, Burkina Faso, and Cameroon). Agence Française du Développement, Paris.

———. 2005. "Etude d'un fonds de soutien et d'un fonds d'intervention au bénéfice de la filière coton au Burkina Faso." (Study for a price support fund in Burkina Faso). Agence Française de Développement, Paris.

———. 2008a. "Benin Cotton Study." Background paper for Comparative Analysis of Organization and Performance of African Cotton Sectors: Learning from Reform Experience. World Bank, Washington, DC.

———. 2008b. "Burkina Country Study." Background paper for Comparative Analysis of Organization and Performance of African Cotton Sectors: Learning from Reform Experience. World Bank, Washington, DC.

———. 2008c. "Cameroon Country Study." Background paper for Comparative Analysis of Organization and Performance of African Cotton Sectors: Learning from Reform Experience. World Bank, Washington, DC.

———. 2008d. "Mali Country Study." Background paper for Comparative Analysis of Organization and Performance of African Cotton Sectors: Learning from Reform Experience. World Bank, Washington, DC.

Gibbon, P. 1999. "Free Competition without Sustainable Development? Tanzanian Cotton Sector Liberalisation 1994/95 to 1997/98." *Journal of Development Studies* 36 (1): 128–50.

Gilbert, C. 2006. "Trends and Volatility in Agricultural Commodity Prices." In *Agricultural Commodity Markets and Trade: New Approaches to Analyzing Market Structure and Instability*, ed. A. Sarris and D. Hallam, 31–60. Northampton, MA: Food and Agriculture Organization of the United Nations and Edward Elgar.

Glover, D. 1990. "Contract Farming and Outgrower Schemes in East and Southern Africa." *Journal of Agricultural Economics* 41: 303–15.

Goreux, Louis, and John Macrae. 2002. "Liberalizing the Cotton Sector in SSA: Part I, Main Issues." Unpublished, World Bank, Washington, DC.

Gouse, M. 2007. "South Africa: Revealing the Potential and Obstacles, the Private Sector Model and Reaching the Traditional Sector." In *The Gene Revolution: GM Crops and Unequal Development*, ed. S. Fukuda-Parr, 175–95. London: Earthscan.

Hinkle, Laurence, and Peter Montiel. 1999. *Exchange Rate Misalignment: Concepts and Measurement for Developing Countries*. New York: Oxford University Press.

ICAC (International Cotton Advisory Committee). Various issues. *Cotton: Review of the World Situation.*

Jaffee, S. 1994. "Contract Farming in the Shadow of Competitive Markets: The Experience of Kenyan Horticulture." In *Living Under Contract: Contract Farming and Agrarian Transformation in Sub-Saharan Africa*, ed. P. Little and M. Watts. Madison: University of Wisconsin Press.

Jayne, T., L. Rubey, D. Tschirley, M. Mukumbu, M. Chisvo, A. Santos, M. Weber, and P. Diskin. 1995. "Effects of Market Reform on Access to Food by Low-Income Households: Evidence from Four Countries in Eastern and Southern Africa." Department of Agricultural Economics, Michigan State University, East Lansing, MI.

J. M. Consultants. 1995. "*La Compétitivité du Coton dans le Monde.*" Paris : Ministère de la Coopération Française.

Kahneman, Daniel, Jack L. Knetsch, and Richard H. Thaler. 1991. "Anomalies: The Endowment Effect, Loss Aversion, and Status Quo Bias." *The Journal of Economic Perspectives* 5 (1): 193–206.

Leibenstein H. 1966. "Allocative Efficiency vs. 'X-Efficiency.'" *American Economic Review* 56 (3): 392–416.

Lele, Uma, Nicholas Van de Walle, and Mathurin Gbetiobouo. 1989. "Cotton in Africa: An Analysis of Differences in Performance." MADIA Discussion Paper No. 7, World Bank, Washington, DC.

Maro, W., and C. Poulton. 2005. "Tanzania Country Report: 2003/04 Production Season." Report produced for the Competition and Coordination in Cotton Market Systems of Southern and Eastern Africa Project, Wye, Imperial College, London.

National Cotton Council of America. 2002. "Government Roles, Private Actions and the US and World Cotton Market." Paper presented at International Cotton Advisory Committee Conference on Cotton and Global Trade Negotiations, July.

North, D. 1990. *Institutions, Institutional Change, and Economic Performance.* Cambridge, UK: Cambridge University Press.

ODI (Overseas Development Institute). 2004. "Developed Country Cotton Subsidies and Developing Countries: Unravelling the Impacts on Africa." Briefing Paper, Overseas Development Institute, London.

PADECO. 2006. "Strategies for Cotton Sector Development in West and Central Africa." Consultancy report prepared for the World Bank. PADECO, Tokyo.

Pan, X., and C. Valderrama. 2005. "Higher Cotton Price Variability." *Review of the World Situation, January-February*. International Cotton Advisory Committee, Washington, DC.

Poulton, C., P. Gibbon, B. Hanyani-Mlambo, J. Kydd, W. Maro, M. Nylandsted Larsen, A. Osorio, D. Tschirley, and B. Zulu. 2004. "Competition and Coordination in Liberalized African Cotton Market Systems." *World Development* 32 (3): 519–36.

Poulton, C., and B. Hanyani-Mlambo. 2007. "Zimbabwe Country Study." Background paper for *Comparative Analysis of Organization and Performance of African Cotton Sectors: Learning from Reform Experience.* World Bank, Washington, DC.

Poulton, C., and W. Maro. 2007. "Tanzania Country Study." Background paper for Comparative Analysis of Organization and Performance of African Cotton Sectors: Learning from Reform Experience. World Bank, Washington, DC.

Pursell, G., and M. Diop. 1998. "Cotton Policies in Francophone Africa." Unpublished, International Economics Department, World Bank, Washington, DC.

Radetzki, Marian. 1985. "Effects of a Dollar Appreciation on Dollar Prices in International Commodity Markets." *Resources Policy* 11 (3): 158–9.

Sarris, A. 2000. "Has World Cereal Market Instability Increased?" *Food Policy* 25 (2): 337–50.

Stringfellow, Rachel. 1996. "Smallholder Outgrower Schemes in Zambia." Research report completed under ODA crops post-harvest programme, project no. AO439. Natural Resources Institute, Chatham, England.

Thirtle, C., L. Beyers, Y. Ismael, and J. Piesse. 2003. "Can GM Technologies Help the Poor? The Impact of Bt Cotton in Makhathini Flats, KwaZulu-Natal." *World Development* 31: 717–32.

Tollens, E., and C. Gilbert. 2003. "Does Market Liberalisation Jeopardise Export Quality? Cameroonian Cocoa, 1988–2000." *Journal of African Economies* 12 (3): 303–42.

Tschirley, D. 2008. "Zambia Country Study." Background paper for Comparative Analysis of Organization and Performance of African Cotton Sectors: Learning from Reform Experience. World Bank, Washington, DC.

Tschirley, David, and Stephen Kabwe. 2007a. "Urgent Need for Effective Public-Private Coordination in Zambia's Cotton Sector: Deliberations on the Cotton Act." Policy Synthesis No. 21. Food Security Research Project, Lusaka. http://www.aec.msu.edu/agecon/fs2/zambia/index.htm.

———. 2007b. "Zambia Country Study." Background paper for Comparative Analysis of Organization and Performance of African Cotton Sectors: Learning from Reform Experience. World Bank, Washington, DC.

Tschirley, David, Colin Poulton, and Duncan Boughton. 2008. "The Many Paths of Cotton Sector Reform in East and Southern Africa: Lessons from a Decade of Experience." In *Hanging by a Thread: Cotton, Globalization, and Poverty in Africa*, ed. William G. Moseley and Leslie C. Gray, 123–58. Ohio University Press.

Tschirley, David, Ballard Zulu, and James Shaffer. 2004. "Cotton in Zambia: An Assessment of its Organization, Performance, Current Policy Initiatives, and Challenges for the Future." Working Paper 10, Zambia Food Security Research Project, Lusaka. http://www.aec.msu.edu/agecon/fs2/zambia/research.htm.

United States. 1990. Cotton Program: The Marketing Loan Has Not Worked: Report to the Honorable David Pryor, United States Senate. Washington, DC. The Office.

———. 1995. Cotton Program: Costly and Complex Government Program Needs to Be Reassessed: Report to the Honorable Richard K. Armey, House of Representatives. Washington, DC. The Office.

Valdès, A., and W. Foster. 2003. "Special Safeguards for Developing Countries: A Proposal for WTO Negotiations." *World Trade Review* 2 (1): 5–31.

Williamson, O. E. 1985. *The Economic Institutions of Capitalism*. New York: The Free Press.

World Bank. 2007. Strategies for Cotton in West and Central Africa: Enhancing Competitiveness in the "Cotton-4" - Benin, Burkina Faso, Chad, and Mali. Washington, DC: World Bank.

Yanggen, David, Valerie Kelly, Thomas Reardon, and Anwar Naseem. 1998. "Incentives for Fertilizer Use in Sub-Saharan Africa: A Review of Empirical Evidence on Fertilizer Response and Profitability." MSU International Development Working Paper No. 70, Michigan State University, East Lansing, MI.